CONCEIVING RISK,

BEARING RESPONSIBILITY

CONCEIVING RISK,
BEARING RESPONSIBILITY

Fetal Alcohol Syndrome &
the Diagnosis of Moral Disorder

Elizabeth M. Armstrong

THE JOHNS HOPKINS UNIVERSITY PRESS
Baltimore

Johns Hopkins Paperback edition, 2008
9 8 7 6 5 4 3 2 1

An earlier version of chapter 3 was previously published as "Diagnosing Moral Disorder: The Discovery and Evolution of Fetal Alcohol Syndrome" in *Social Science and Medicine* 47, no. 12 (1998): 2025–42, and is used with permission from Elsevier Science.

The Johns Hopkins University Press
2715 North Charles Street
Baltimore, Maryland 21218-4363
www.press.jhu.edu

The Library of Congress has catalogued the hardcover edition of this book as follows:

Armstrong, Elizabeth M., 1967–
     Conceiving risk, bearing responsibility : fetal alcohol syndrome and the diagnosis of moral disorder / Elizabeth M. Armstrong.
          p. cm.
Includes bibliographical references and index.
     ISBN 0-8018-7345-2 (hardcover : alk. paper)
     1. Fetal alcohol syndrome—Social aspects—United States.   2. Fetal alcohol syndrome—United States—History.   I. Title.
     RG629.F45 A767   2003
     618.3'268—dc21   2002152399

ISBN-13: 978-0-8018-9108-3
ISBN-10: 0-8018-9108-6

A catalog record for this book is available from the British Library.

In memory of Bernard F. Armstrong Sr.,

my beloved Pappy,

who taught me faith, hope, and charity,

which turned out to be all I really needed to know

CONTENTS

ILLUSTRATIONS AND TABLES

ACKNOWLEDGMENTS

For a long time now, I have believed that our accomplishments are never ours alone but are shared with those around us who, in invisible and all too often unacknowledged ways, make it possible for us to carry out our work and live our lives. Writing this book has only convinced me more deeply of that truth. I could never have completed it without the help — material, spiritual, and intellectual — of many people. One of the joys of finishing the book is the opportunity to enumerate here those who share in its achievement with me. Needless to say, I alone bear the burden of its faults.

To my mentors at the University of Pennsylvania, I owe an enduring debt of gratitude. I am grateful to Sam Preston for his time and attention, especially when administrative pressures loomed large. I am also grateful to Charles Rosenberg. Even though I was the one spending hours in the library with Index Medicus, sometimes it seemed he knew more about my subject than I did, his knowledge of American medical and social history is so vast; I thank him for sharing it with me and for helping me to become at least a part-time historian. Renée Fox is of the rarest and finest class of scholars. Her warmth, generosity, and grace have long sustained and inspired me. She is quite simply among the very best teachers I know. Chuck Bosk is one of the smartest people I know; he always seemed to see clearly and instantly the unpolished gems in my findings. Though he may be surprised to hear it, he was a kind of safe harbor from the maelstrom of research and writing, as well as a superb guide through the thickets of sociological theory and practice. I continue to be comforted by his astute observation that "there are an infinite number of mistakes you can make for the first time." To my wry regret, I've yet to disprove this axiom!

Of my book group, Sara Curran, Amy Borovoy, and Carolyn Rouse, two things are indisputably true. First, without their intellectual and emotional companionship, I never would have made it through the long and arduous revision process. Second, by virtue of their brilliant insights, gentle criticisms, and trenchant suggestions, this book became an immeasurably better one than I could have written on my own. They helped to catalyze a clarity in my ideas and in my prose that was but dimly present before. They also gave me a reason to keep writing when the task seemed impossible. This is the sort of debt one can never repay. Thank you, Carolyn. Thank you, Amy. Thank you, Sara.

Without my siblings, Trip, Tim, Katie, and Lauren, whom I count among my greatest blessings, this undertaking — not to mention my entire life — would have been far lonelier. They have provided many occasions for escape, laughter, love, and reminders that work isn't the most important thing in the world. My mother, Honey Armstrong, transcribed my interviews and made that onerous task a labor of love, as she has made her whole life.

Many friends and colleagues have contributed to this book. My Philadelphia crew, especially Meredith Kelsey, Peter McGrath, Dan McGrath, and Ellen Foley, were present throughout the initial incarnation of the book as a doctoral dissertation. Their company — not to mention their cocktail-mixing skills — made graduate school bearable. Cynthia Harper and Charlotte Ellertson have wholeheartedly shared tribulations and triumphs alike from afar. Their friendship is like pure sunshine. Jennifer Hirsch has been a tried and true friend as well as a trusted colleague. Peter Jacobson, Peter Conrad, and Renée Anspach provided encouragement and helpful insights along the way. Paula Lantz is everything I aspire to be — a rigorous scholar, a dedicated teacher, a warm and funny friend. James Trussell has been unwavering in his support. Many other friends have sustained and encouraged me along the way, especially Alisa Pascale, Maureen Phipps, and Judith Levine.

I am grateful to Jacqueline Wehmueller and Carol Pech at the Johns Hopkins University Press, who both seem blessed with equal measures of keen editorial acumen and forbearance. They suffered patiently through many tortured revisions and long delays with greater grace than I am sure I deserved. I have had the good fortune to pursue this research and write this book with the generous institutional and financial support of the University of Pennsylvania, the University of Michigan, Princeton University, the National Science Foundation, the Francis Clark Wood Institute for the History of Medicine at the College of Physicians of Philadelphia, and the Robert Wood Johnson Foundation. The marvelous library of the College of Physicians of Philadelphia was an invaluable source of the materials described in Chapter 2. I am indebted as well to Dr. William Beck, Dr. Marjorie Bowman, and Dr. Stephen Ludwig, who helped facilitate my physician interviews, and especially to the thirty doctors who shared with me their knowledge, wisdom, and experience. Judie Miller helped me to prepare the manuscript with her characteristic patience and good cheer.

Finally, my husband, Chris Weiss, suffered the toll that writing this book took in so many ways. That he bore this unsought burden with immense patience, with an unfailing sense of humor, with an infinite supply of encouraging words, and, most especially, with an inexhaustible faith that I could do it is a testament to why I married him in the first place. Thanks, teammate.

CONCEIVING RISK,
BEARING RESPONSIBILITY

CHAPTER I

CONCEIVING RISK

*I think it's high time for women to be held accountable when
they inflict injuries on their unborn children.*
— Joan Korb, Racine County Assistant District Attorney

On the night of March 16, 1996, Deborah Zimmerman was brought into the
emergency room at St. Luke's Hospital in Racine, Wisconsin. The thirty-four-
year-old woman was angry, belligerent, and very drunk. Her blood alcohol
level was above 0.3 percent, more than three times the state's legal limit of intoxication. Zimmerman was also nine months pregnant and just a week shy of
her due date. Doctors at the hospital wanted to admit her and deliver her baby
that night. At first she refused. According to a nurse's testimony, the expectant
mother said, "I'm just going to go home and keep drinking and drink myself
to death, and I'm going to kill this thing because I don't want it anyways."[1] The
doctors eventually prevailed; Zimmerman consented to a cesarean section,
and her daughter was born later that night. The infant, named Megan, weighed
four pounds, six ounces, and had a blood alcohol level of 0.199 percent. Because the baby had a low birthweight and had mild facial abnormalities, the
attending neonatologist suspected she might have a condition he characterized
as "fetal alcohol effect." After a few weeks in the hospital, Megan was dis-
charged to the care of a foster family.[2]

Three months later, Deborah Zimmerman was charged with attempted
murder and reckless endangerment. The latter charge stemmed from the suspected diagnosis of "fetal alcohol effect" at the time of Megan's birth. Although
Zimmerman was not the first woman in the United States to be charged with
harming her unborn child through her substance use, she was one of the first
to face such charges based on drinking — a legal activity — rather than on illicit
drug use. And she *was* the first to face a homicide statute. Joan Korb, the as-

sistant district attorney who pressed the charges against Zimmerman, explained her decision to prosecute: "I think it's high time for women to be held accountable when they inflict injuries on their unborn children."[3]

Zimmerman spent three years in jail while her case made its way through Wisconsin's judicial system. The charges brought by the district attorney against Zimmerman withstood court scrutiny at several levels before Wisconsin's Second District Court of Appeals ruled in May 1999 that Zimmerman could not be charged with attempted homicide because Wisconsin statutory law does not recognize an unborn fetus as a human being.[4]

Deborah Zimmerman's story seems both extraordinary and typical. It is extraordinary that a patient's explicit suicide threat, made in the presence of doctors and nurses, generated no psychiatric consultation, evaluation, or treatment. It is extraordinary that a pregnant woman could be charged with attempting to murder the fetus she carried to term. It is as if no one took seriously the first part of her statement about wanting to kill herself, but society held her responsible for the second part of her statement. It is extraordinary that a woman could be prosecuted for a behavior — drinking — that may sometimes be stigmatized in our society but is just as often glamorized and is in no way illegal. And yet, viewed another way, the story seems all too typical. In the decade before Zimmerman's prosecution, hundreds of women across the country faced charges for their illicit drug use during pregnancy — charges ranging from child abuse, to delivery of drugs to a minor, to assault with a deadly weapon. The ideas about prenatal risk and maternal responsibility that ani-mate the charges against Zimmerman and hundreds of other women reflect growing social unease about "fetal harm," particularly when it is purportedly inflicted by the pregnant woman herself. Thus, Deborah Zimmerman's case typifies and crystallizes a growing social conflict over reproduction.

What does it mean to "hold accountable" women like Deborah Zimmerman? That alcohol is dangerous during pregnancy and that women ought to be responsible for the outcomes of their pregnancies are regarded as well-known "facts" in American society today. Pregnant women who have even a single drink routinely face harassment, social stigma, and openly voiced reproach from both social intimates and total strangers. Because of state and local ordinances mandating point-of-purchase warnings, women and men in many places are reminded of these facts every time they go to a bar, a restaurant, or a liquor store. Everywhere, we are all reminded of these facts when we glance at the warning label that the federal government mandates for every can and bottle of beer, wine, and liquor sold in the United States. Yet how did these facts come to be accepted as knowledge — "what everybody knows"? And how does this knowledge shape the way we think about and understand risk and responsibility — how we decide who should be "held accountable" and why? Pregnant women in the United States are more likely to suffer domestic violence, for example, than to drink excessively during pregnancy. Yet public policy acknowledges and acts on one risk but not on the other. In other words, how do we know what we know? And what are the consequences of that knowledge, particularly for how we think about a set of very important relationships: the relationships between women who are pregnant and the rest of us, between women and their unborn children, between doctors and their patients, and between individuals and society? How is our concept of responsibility shaped by our changing ideas about motherhood, the nature of the fetus, and the relationship between the two? These are the questions this book takes on, using fetal alcohol syndrome and the problem of drinking during pregnancy to examine the assumed relationship between somatic disorder and social disorder, the ways in which social problems are medicalized and thus individualized, and the intertwining of health and morality that characterizes American society.

Although Megan Zimmerman was born with alcohol in her blood, it was not this that led the doctor to suspect a diagnosis of "fetal alcohol effect." The terms *fetal alcohol syndrome* and *fetal alcohol effect* are not used to describe a baby "born drunk." Rather, *fetal alcohol syndrome* (FAS) describes a set of birth defects associated with prenatal alcohol exposure; *fetal alcohol effect* (FAE) is sometimes used to denote a less severe manifestation of the same symptoms. A medical diagnosis of FAS requires that four distinct criteria be met (see Table 1). First, there must be confirmed maternal exposure to alcohol. The mother must exhibit a pattern of excessive alcohol intake, characterized by substantial, regular drinking or heavy, episodic drinking (sometimes called binge drinking). Second, the child must demonstrate certain craniofacial anomalies. These may include microcephaly (small head size), short palpebral fissures (which give the eyes a rounded look), epicanthal folds in the eyelids, ptosis (droopiness) of the eyes, flattened midface, flat or indistinct philtrum (the groove between nose and upper lip), and a thin upper lip. Third, the child must have prenatal or postnatal growth retardation. Fourth, the child must suffer central nervous system anomalies, such as structural brain abnormalities, neurological impairment (including poor hand-eye coordination), impaired fine motor skills, poor tandem gait, hearing loss, or mental retardation.[5]

Table 1. Diagnostic Criteria for Fetal Alcohol Syndrome

Category	Criteria
Confirmed maternal alcohol exposure	A pattern of excessive intake characterized by substantial, regular intake or heavy episodic drinking. Evidence of this pattern may include frequent episodes of intoxication, development of tolerance or withdrawal, social problems related to drinking, legal problems related to drinking, engaging in physically hazardous behavior while drinking, or alcohol-related medical problems such as hepatic disease.
Evidence of characteristic pattern of facial anomalies	Includes features such as short palpebral fissures and abnormalities in the premaxillary zone (e.g., flat upper lip, flattened philtrum, and flat midface).
Evidence of growth retardation	At least one of the following: • low birthweight for gestational age • decelerating weight over time not due to malnutrition • disproportionately low weight for height
Evidence of CNS neurodevelopmental abnormalities	At least one of the following: • decreased cranial size at birth • structural brain abnormalities • neurological hard or soft signs

Source: Kathleen Stratton, Cynthia Howe, and Frederick Battaglia, eds., *Fetal Alcohol Syndrome: Diagnosis, Epidemiology, Prevention, and Treatment* (Washington, D.C.: National Academies Press, 1996), © 1996 National Academy of Sciences.

The purpose of this book is not to question the existence of fetal alcohol syndrome. There is good evidence for a recognizable syndrome of severe birth defects associated with high levels of alcohol exposure in utero; it is not my aim to debunk the diagnosis of fetal alcohol syndrome. Rather, the purpose is to examine more clearly what we know and do not know about drinking during pregnancy and the harm it causes. For example, fewer than 5 percent of the babies born to women who drink heavily during pregnancy are affected by FAS; how can we reconcile this fact with claims that all pregnant women must avoid alcohol? Indeed, there is little evidence substantiating such diagnoses as "fetal alcohol effect," "alcohol-related birth defects," "partial FAS," or "alcohol-related neurodevelopmental disorder." Moreover, there are substantial gaps in our understandings of FAS itself. It is precisely these gaps that this book examines. The question is not whether fetal alcohol syndrome is real or not. The pertinent questions are, rather, how well do we really understand the etiology of alcohol teratogenesis, and what is the relationship between alcohol exposure in pregnancy and birth outcome? What is the distribution of FAS in the population? Who is really at risk? This book explores how knowledge gets transformed in the public imagination and then translated into public policy. Why does medical knowledge become codified even in the face of uncertainty? The answer lies with the power of medical knowledge to capture social anxieties.

The Institute of Medicine of the National Academy of Sciences estimates that some two thousand to twelve thousand children are born with FAS every year in the United States. As this wide range indicates, the epidemiology of fetal alcohol syndrome is murky. Estimates of the incidence of FAS made by different researchers vary widely, ranging from 1 to 10 cases per 10,000 births for the United States overall, with even greater variation among population subgroups.[6] The Birth Defects Monitoring Program (BDMP) of the federal Centers for Disease Control and Prevention identified rates ranging from 0.9 cases per 10,000 births among whites to 6 cases per 10,000 among blacks to 29.9 cases per 10,000 births among American Indians.[7] Because the syndrome is notoriously difficult to diagnose and because there is no biological marker that can confirm the diagnosis, FAS is subject to ascertainment bias — that is, doctors may be predisposed to see it more in some groups than in others.[8] Moreover, the craniofacial abnormalities, which are the most distinctive element of the syndrome, may change with age, becoming less noticeable, and some of them may be confounded by racial phenotype.[9]

Despite the considerable ambiguity surrounding the diagnosis and estimates of the incidence of alcohol-affected births, many observers claim that their incidence is much higher than reported. They use such categories as fetal alcohol effect (the label Megan Zimmerman received), alcohol-related birth defects (ARBD), partial FAS, and alcohol-related neurodevelopmental disorder (ARND) to capture this wider spectrum. The last of these is a particularly diffuse diagnosis. The Institute of Medicine defines ARND as "evidence of a complex pattern of behavior or cognitive abnormalities that are inconsistent with developmental level and cannot be explained by familial background or environment alone, such as learning difficulties; deficits in school performance; poor impulse control; problems in social perception; deficits in higher level receptive and expressive language; poor capacity for abstraction or metacognition; specific deficits in mathematical skills; or problems in memory, attention or judgment" in conjunction with "a history of maternal alcohol exposure."[10] This broad definition could be applied to a large number of children. However, as the Institute of Medicine acknowledges, "at present, there are no universally applied diagnostic criteria or instrument(s) for the diagnosis of FAS, ARBD, and ARND."[11] Since what precisely constitutes a diagnosis of FAS, FAE, ARBD, ARND, or any of these other labels may vary from study to study, incidence rates are neither definitive nor readily comparable across studies.

Although FAS and its kindred syndromes are often presented as clearly established diagnostic paradigms, and indeed are often identified as the "leading preventable cause of birth defects in the United States," considerable uncertainty pervades our understanding of the relationship between alcohol and reproductive outcome. One large area of uncertainty, as described above and more in depth in Chapter 5, is the epidemiology of the syndrome. Another is its etiology: what *causes* fetal alcohol syndrome? As the name itself implies, alcohol is assumed to be the teratogenic agent at work, but there is much we do not yet understand. The relationship between alcohol and outcome is riddled with puzzles and paradoxes. We do not know how much alcohol causes what kinds of birth defects. Although some women who drink very heavily during pregnancy have babies who suffer the set of symptoms labeled as FAS, there is no consistent evidence that all women who drink heavily during pregnancy will have babies with fetal alcohol syndrome or that conditions with such umbrella diagnoses as FAE and ARND exist at all. Both in the public imagination and in much of the medical literature, it is assumed that if heavy alcohol exposure causes severe birth defects, then lesser levels of exposure must cause more moderate effects. The fact is we simply do not know; however, there is no consistent, reliable evidence from prospective studies of women who drink during pregnancy to indicate that alcohol categorically affects fetal development regardless of level of exposure or timing of exposure, or absent other factors. What is important to stress is that while ample evidence indicates that *heavy* prenatal alcohol exposure can be teratogenic in some pregnancies, other types of evidence suggest that the number of fetuses exposed and affected is small. Drinking during pregnancy is not common. There is a discordance between evidence and imagination.

First, we know that FAS is not an "equal opportunity" birth defect.[12] Fetal alcohol syndrome is highly correlated with smoking, poverty, malnutrition, high parity, and advanced maternal age. Each of these factors alone increases the risk of an adverse birth outcome; we do not understand their synergistic and combined effects. Alcohol exposure in utero appears to be a necessary condition for the development of FAS, but whether it is also sufficient we simply do not know. It may be that the teratogenic effect of alcohol is expressed only in the presence of predisposing "permissive" sociobehavioral factors, such as smoking, binge drinking, or poverty, that can produce certain "provocative" biological factors, such as undernutrition, maternal stress, or exposure to environmental pollutants, that increase fetal vulnerability to alcohol's teratogenic effect.[13]

Second, despite evidence to suggest that FAS may have a genetic component, there has been surprisingly little research exploring this hypothesis. The evidence comes in two forms. One is from studies of twins: the medical literature has reported cases of dizygotic (fraternal) twins, in which one twin is affected with FAS and the other is not.[14] These infants shared the same uterine environment throughout pregnancy, so each would have been equally exposed to alcohol in utero, but the affected twin may have a genetic anomaly that precipitates the teratogenic effect of alcohol. One hypothesis is that fetuses or women who lack a crucial enzyme (alcohol dehydrogenase) for the breakdown of ethanol in the body are at vastly increased risk of FAS. The absence of this enzyme could be genetically determined.[15] The other hint of a possible genetic component is the well-documented phenomenon that, once a woman has a child with FAS, the odds of her having a later child with FAS are greatly magnified. In other words, fetal alcohol syndrome appears to cluster among siblings.[16]

Third, even among women who are alcoholics, only a small number have babies who are affected. Various studies have shown that not all women who drink during pregnancy, even those who drink heavily, have children with FAS. In fact, only about 5 percent of children born to alcoholic women have FAS.[17] This finding has led some researchers to speculate that alcohol — or its primary ingredient, ethanol — may not be the teratogenic agent at work. It may be that some by-product of ethanol breakdown in the body — acetaldehyde perhaps — causes the teratogenesis observed with alcohol exposure. Or it may be that alcohol works in concert with other factors to affect fetal development.

Finally, we do know that drinking is not universal among women of childbearing age: only half of women aged eighteen to forty-four report any drinking at all.[18] Drinking is even less common among pregnant women; fewer than one in five women (16 percent) report drinking during pregnancy, and most of these women report consuming fewer than seven drinks per week, a level that probably does not pose a risk of FAS.[19] Most studies, as Chapter 5 will show, suggest that heavy prenatal drinking is concentrated among a small number of women in the United States.

Given the uncertainties surrounding the epidemiology and etiology of FAS and its kindred syndromes, we might wonder why these "facts" — that alcohol is dangerous in pregnancy and that all women should abstain from drinking to ensure healthy babies — came to be accepted as knowledge. How did drinking during pregnancy come to be considered such a pervasive social problem, despite the weak empirical foundation? The answer lies at the intersection of medical knowledge, social problems, gender, and the body, and in the uncertainty of our knowledge about the epidemiology and etiology of fetal alcohol syndrome.

Uncertainty pervades human life. Sociologists have long observed that uncertainty is a core element of modern medicine, where it lies at the underside of all knowledge and practice.[20] In the face of uncertainty, diagnosis — the codification of medical knowledge in discrete disease categories — serves to impose order on the inherently disorderly human experience of illness, disease, and disability. Diagnosis is the necessary foundation for therapeutic intervention; it is at once a response to a specific clinical dilemma — what ails this patient? — and a response to a more general epistemological question — how should we interpret the myriad conditions afflicting the human body and impairing human functioning? Diagnosis thus implies both etiology and treatment. Etiology in turn intimates risk; treatment implies responsibility. However, diagnostic categories like FAS and FAE may also be used to exorcise uncertainty and ambiguity. The medical diagnosis of FAS, I will argue in Chapter 4, represents an attempt to manage uncertainty and to create explanation in the social as well as the clinical realm. The stories we tell about illness, the scientific theories we hypothesize, the explanations we concoct, both reflect and reinforce the social order. Doctors "create illness as social meaning."[21] Embedded within medical conceptions of fetal alcohol syndrome, as within other explanations for illness, are "deeply held moral convictions about the nature of risk and responsibility for disease."[22] Medicine is an arena in which we increasingly express and wrestle with questions about the kind of world we live in and the kind of world we want to live in. Sociological and historical observers have long noted the ways in which medicine functions as a form of social control. Talcott Parsons, the first sociologist to turn sustained attention to the discipline of medicine, characterized medicine's task as one of "coping with disturbances."[23] By this he meant not only the physiological disturbances brought on by disease but the perturbations of the social order or social norms as well. In the process of classifying physiological disorder, medicine also functions to create and maintain social order. We have always sought to impose moral meaning on illness, "to constrain misfortune in reassuring frameworks of meaning."[24] Ideas about risk and danger in the realm of health serve as "useful tools for demanding moral behavior."[25]

Indeed, the uncertainty surrounding alcohol and reproduction creates a space in which medical knowledge *and* social relationships are redefined. The borders of that space are demarcated by our social determinants of risk. Both medicine and American society are risk averse; we use medicalization as a strategy to control and manage risk. In the case of FAS, uncertainty about the physiological disorder — what causes it? who is affected by it? — intersects with uncertainty about the social order. What should the role of women in society be? What are our obligations to the next generation? Can we guarantee our collective future? Medical knowledge is significant because doctors and medical practice are often at the nexus of these uncertainties. The discovery of FAS was made possible by a shift in medical knowledge. Once FAS was "discovered" in 1973, medical knowledge changed. The notion of discovery supposes that FAS existed all along and was suddenly found or recognized by observant doctors. Yet, as we will see in Chapters 2, 3, and 6, medical knowledge itself is the product of social conditions and historical change. To be more precise, the way doctors think about drinking and pregnancy has changed dramatically over the last 150 years. This creation of a new strand of medical knowledge centered around the fetus has had profound consequences beyond the realm of obstetrics, because the transformation in medical knowledge both reflected and precipitated broader social change.

By changing our perceptions of risk and responsibility, medical knowledge in the form of the diagnosis of FAS helps to *redefine* the bodily relationship between the woman and the fetus. The crucial movement in this redefinition is from thinking of the woman and the fetus as a single entity to thinking of the woman and the fetus as two separate individuals. From thinking of pregnant as something a woman is to regarding pregnancy as something she *carries*. Once we conceive of the pregnant woman and the fetus as two separate individuals rather than as one, it becomes possible to think of the woman and her fetus as potential antagonists. The knowledge that undergirds the clinical diagnosis of fetal alcohol syndrome helps us to think in terms of "maternal-fetal conflict" — for Joan Korb to accuse a suicidal Deborah Zimmerman of attempted homicide and to hold her accountable for harm to an "other." Diagnosis shapes social norms, what we think is right and orderly. Thus, in redefining a bodily relationship, medical knowledge about fetal alcohol syndrome redefines a set of *social* relationships as well. The body mediates social relations and becomes a site for contesting social order.[26]

Controlling women's bodies is one strategy for managing risk, though as Chapter 6 argues, it does not appear to be a particularly effective one. This reconception of the meaning, nature, and status of pregnancy makes the woman a fetal vessel — she is merely the carrier of the fetus. In George Annas's memorable words, the pregnant woman becomes a mere "fetal container."[27] This process threatens to dehumanize and objectify the pregnant woman.

Despite the firmness of the boundaries drawn by medical knowledge in the form of the diagnosis of FAS, the medical knowledge itself, the practice and application of the diagnosis, is far more indeterminate. Chapter 4 examines the implication of this fuzziness for how doctors think about FAS — and more broadly about their responsibility to their patients. Doctors utilize or refuse diagnoses as a way to make an impact on social problems.

A powerful dialectic process is at work as well. Medical knowledge does not evolve in a vacuum; it is situated in society and in history. When uncertainty prevails, doctors may be especially prone to drawing on current social metaphors and ideals in their descriptions and explanations of illness, thus reifying those norms. Part of what prompted the medical revision of pregnancy risk was the shifting boundaries of gender in this period. FAS did not just happen to be discovered in 1973; rather, shifting social norms, beliefs, and relationships became reordered in such a way that FAS "appeared," just as when a piece of stained glass is in changing light, a deepening or lightening of background color suddenly brings an unnoticed detail to the foreground. In the case of FAS, medical ideas arose out of cultural ferment over gender and motherhood, but they also further leavened that ferment.

The uncertainty of medical knowledge about the physiological disorder of FAS is connected to deeper uncertainty about the sources of and appropriate "treatment" for social disorder. In fact, the diagnosis of FAS crystallizes a particular paradox of care that doctors face when dealing with disorders that are characterized as the consequence of deliberate human behavior as much as physiological dysfunction and that are without cure. Social problems, like drinking, that manifest themselves in patients' bodies present dilemmas that are not easily resolved by physicians, forcing them to demarcate the boundaries of their professional responsibilities and to articulate the limits of their professional competence. In patients such as children with FAS or women who, like Deborah Zimmerman, drink heavily, perhaps even uncontrollably, during pregnancy, doctors encounter the weaknesses and sicknesses not only of individual bodies but of the body of society. How do they reconcile the relationship between these two kinds of illness, the individual and the social? The dilemma presents an acute paradox of care: often doctors cannot heal the individual patient's body without treating the social one as well, yet to do so is well beyond their professional training and mandate. One response is to internalize social order.[28] "The model of order is to be found inside the individual."[29] Fetal alcohol syndrome exemplifies this predicament. Physicians' responses to the diagnosis, as Chapter 4 shows, thus reveal not only their technical skills as clinicians but the kind of moral framework they believe buttresses their profession and the social obligations it entails.

While doctors may struggle to manage both kinds of illness — the individual and the social — our wider social policies with respect to drinking during pregnancy tend to ignore the social dimension and emphasize the individual. Moreover, policies such as the U.S. surgeon general's warning that pregnant women should abstain from alcohol and the federal government's alcoholic beverage labeling law abjure uncertainty. Such policies define *all* women as being *equally* at risk, even though the clinical and epidemiological portraits of the syndrome exhibit considerably more complexity, as we will see in Chapters 4 and 5. And such policies assume that the "solution" to FAS is to tell women not to drink. Yet not only do these policies fail to achieve their intended goal, they also ignore the social context in which FAS occurs. The construction of the diagnosis of FAS reflects the emergence of an epidemiological paradigm emphasizing risk factors rooted in individual behaviors and predilections; indeed, it epitomizes our modern American tendency to locate explanations for disease at the individual — whether behavioral or biological — level rather than to see disease as the consequence of social conditions.[30] Americans' tendency to attribute social ills to individuals' biology rather than to society mirrors our propensity to look for solutions to problems in medical technology, cures, treatments, and changes in individual behavior or "lifestyle." Biologizing problems also individualizes them — taking them out of the realm of society and embedding them in the bodies of individual women, children, and men. This process of medicalization and subsequent desocialization is a powerful tool for maintaining the status quo, since it reinforces the belief that individual, not social, change is called for.[31] As we will see in Chapter 6, the response to FAS and to drinking during pregnancy is very different in the United States than elsewhere in the world. The prosecution of Deborah Zimmerman is but an extreme example of the kind of zero-tolerance, punitive approach that often characterizes the American policy response to social disorder.

Medicine provides an arena in which society can sift through and express moral ambiguities. But medical knowledge is also of particular interest, because it has consequences far beyond the realm of clinical medicine. "Like all other human action, medicine is active interpretive work through which a particular social reality is constituted by diagnosed illness and prescribed treatments."[32] It diffuses outward to society; it informs social policy and it informs individual choices. Medical knowledge about FAS colors a far wider swath of social interactions than merely that between doctor and patient. Modern medical ideas about the relationship between drinking and reproductive outcome affect the interactions between wives and husbands, between liquor store clerks and customers, between waiters or bartenders and their patrons, even between prosecutors and the women they target as defendants, as the Zimmerman case shows. Perhaps most significantly and certainly most intimately, the idea of FAS reconfigures the formerly sacrosanct bond between the pregnant woman and the fetus.

This book traces the evolution of medical knowledge and social thought about the effect of alcohol on reproduction, from nineteenth-century debates about drinking and heredity to the modern diagnosis of fetal alcohol syndrome, arguing that throughout the last two centuries, medical diagnosis represents an attempt to manage uncertainty and to create explanation in the social as well as the clinical realm. In particular, the modern diagnosis of FAS reveals how medical knowledge shapes social knowledge and, conversely, how social thought infuses and directs medical inquiry. The book delineates the continuities and discontinuities in the way American medicine has viewed the relationship between alcohol and reproduction over the last 150 years. Throughout this period, medical knowledge has been powerfully influenced by social context, prevailing morality, and social norms, particularly those concerning gender. An examination of historical evidence shows that medical assessments of alcohol's effects on offspring have always reflected broader social preoccupations, though couched in the clinical language of diagnosis. Moreover, the categories that medicine has created to explain these effects have in turn diffused into society as well as into clinical judgment and practice. Just as significantly, medical understanding has had an enormous impact not only on public understandings of risk in pregnancy and public policy but on the private experience of pregnancy for millions of women.

We cannot begin to make sense of Deborah Zimmerman's story in 1996 without understanding the history of the diagnosis of fetal alcohol syndrome, the evolution of the notion of maternal-fetal conflict, and the social, legal, and political push to hold women alone accountable for reproductive outcomes. Indeed, in order to understand the present-day configuration of drinking during pregnancy as a "major public health problem," we need to step back to 1973, when the diagnosis of FAS first appeared in the medical literature. Why was FAS discovered in 1973? Although Chapters 3 and 6 offer more detailed answers, I will briefly consider the matter here. Certainly women had been drinking during pregnancy for centuries. Yet the relationship between prenatal alcohol exposure and adverse outcome became visible to physicians in 1973 in a new way. As Chapter 6 reveals, the diagnosis emerged at the confluence of a number of cultural currents; fetal alcohol syndrome appeared when and where it did because social and historical conditions facilitated its recognition.

Perhaps the most significant of these cultural currents was the reproductive revolution unfolding at that time, beginning with the Food and Drug Administration's approval of the oral contraceptive pill in 1960 and continuing with the legalization of abortion in 1973. The reproductive revolution profoundly reordered relationships between women and men, between women and doctors, and between women and society as a whole. And it heralded a transformation in how people thought about conception, pregnancy, childbirth, and even parenthood. Central to this transformation was the notion of control: women could control the timing, pace, and level of their fertility. In addition, the entire process of reproduction became commodified in this era; both the materials of gestation — eggs, sperm, the uterus — and its result — the child — came to be seen as "products" that were available on the open market and subject to the same quality control applied to manufactured products.[33] As families became smaller, parents developed "the perfect child mentality."[34] In fact, this cultural shift in emphasis from quantity of children to quality of children began much earlier, in the nineteenth century, with the sacralization of childhood as a separate and distinct phase of life. However, the shift greatly intensified in the twentieth century as this sacralization of childhood was popularized and spread from the upper to the lower classes.[35] The move towards fetal personhood extends this process by making gestation a separate and distinct phase of childhood. Thus, all aspects of reproduction became subject to control along two dimensions: quantity and quality.

At the same time, new medical technology was reconfiguring medical and lay knowledge of pregnancy. Fetal medicine was coming into its own as a field of specialization; the fetus was becoming a patient in its own right.[36] Advances in medical technology — namely, ultrasonography and fetal monitoring — for the first time enabled doctors, pregnant women, and the public to see (literally) the pregnant woman and the fetus as two distinct individuals rather than one. Prenatal technology today makes the mother "transparent" — it erases her in the act of illuminating the fetus. Once this fundamental detachment had occurred, the relationship between these two "individuals" became increasingly problematic, estranged, and embattled. The growing separation of mother and fetus took place along two dimensions, as maternal-fetal antagonism was superimposed on the increasing individualization taking place more generally in society. Concurrently, there was a growing tendency in medicine, law, and society at large to see the pregnant woman and the fetus as not only separate but at odds, their individual welfares as distinct and oppositional, rather than mutual. It is no coincidence that the diagnosis of FAS emerged in 1973, the year that the Roe v. Wade decision legalized abortion in the United States; both events express the changing perception of the relationship between woman and fetus, as well as a growing social desire to control reproductive outcomes and belief in the power to do so. By giving women the option to end an unwanted pregnancy, legalized — and thus socially sanctioned — abortion makes transparent the potential conflict between the fetus and the pregnant woman.

Although this perception of a conflict between the pregnant woman and the fetus may have been new in 1973, the diagnosis of FAS also expressed much older notions about pregnancy and maternal responsibility. Indeed, just as we cannot understand the discovery of FAS without taking into account the wider social landscape, we cannot make sense of this configuration of medical knowledge without taking a deeper view of history, too. Although it was not until 1973 that the modern diagnosis of fetal alcohol syndrome took root in the medical imagination and lexicon, medicine had long pondered the relationship between drinking and reproduction. Plato, for example, observed that "it is not right that procreation should be the work of bodies dissolved by excess of wine."[37] Yet, earlier generations' understanding of that relationship is very different from our current conception of the prenatal risk posed by alcohol exposure in utero. Until the twentieth century, physicians focused not on alcohol's prenatal effect but on the ways in which it distorted heredity. Therefore, drinking by both men and women was believed to pose a threat to the next generation, in the form of such undesirable consequences as a weakened constitution, imbecility, idiocy, epilepsy, and an inherited predisposition to the "drink crave." For centuries, then, doctors had been concerned equally with both male and female bodies "dissolved by excess of wine." The modern diagnosis of fetal alcohol syndrome, in contrast, pinpoints only women's drinking as dangerous, absolving men of responsibility for alcohol-damaged offspring. The modern location of risk — in utero (in other words, in women's bodies) rather than in conception — and the consequent assignment of responsibility to women alone is a significant discontinuity in medical thinking about alcohol and reproduction.

What role does history play in the accretion and erosion of the "facts" that we recognize at any given moment as medical knowledge? Like all knowledge, medical knowledge is a social phenomenon and is not static. But neither is its trajectory necessarily continuous or cumulative — that is to say, medical understandings of FAS today did not necessarily evolve in a straight line from earlier medical knowledge. As with most diseases, medical knowledge about alcohol and reproduction has transmogrified from one generation to the next — in this case, from a focus on the hereditary degenerative properties of alcohol to an emphasis on its teratogenic potential.

Yet tracing the historical roots of fetal alcohol syndrome and understanding the relationship between past and present knowledge is a trickier endeavor than it might at first seem. Our understanding of the relationship between past and present knowledge is complicated by the discontinuities and continuities in the trajectory of medical knowledge. The discontinuities arise from the non-linear nature of medical knowledge. At various points over the last 150 years, doctors have viewed alcohol and reproduction along a spectrum from harmful, even poisonous, to benign, even beneficial.

Chapter 2 examines the "prehistory" of the modern diagnosis of FAS. Although the doctors who discovered the syndrome in 1973 initially heralded their observations about drinking and reproductive outcome as new, medicine had a long history of trying to answer the "question of alcohol and offspring," as one nineteenth-century physician put it.[38] The question of alcohol and offspring might indeed be characterized as an obsession of physicians in this era, particularly those involved in the temperance or eugenics movements. In the nineteenth and early twentieth centuries, as the temperance and eugenics movements flourished in the United States, alcohol was seen variously as a degenerative agent, a racial poison, and a "fool-killer." All these explanations for alcohol's effect rested on heredity as the causal mechanism. Children inherited drunkenness itself from their parents, or alcohol corrupted or poisoned the parents' seed and subsequently the hereditary process. By the end of the nineteenth century, hereditarianism was a powerful explanatory tool applied to just about every aspect of social life.[39] Ideas about heredity took a powerful new form in the early decades of the twentieth century in the theory and practice of eugenics, reflecting a concern with the social order and indeed the speculative and actual manipulation of the very composition of the social body to attain an ideal social order. Each of these movements — temperance and eugenics — was at base about social reproduction, and alcohol played a fundamental role in each. Moreover, these movements, eugenics particularly, relied on control of women's bodies to achieve a particular desired social order. As the temperance and eugenics movements lost appeal in the 1920s and 1930s, and particularly with the repeal of Prohibition in 1933, alcohol vanished from medical thinking about reproduction. During the middle decades of the twentieth century, doctors took a benign view of alcohol in pregnancy, even for a time using it therapeutically. It was as if medicine forgot the earlier accumulated knowledge about alcohol and heredity. Thus, there is no single path along which we can trace medical ideas about alcohol and reproduction.

Nonetheless, there are many continuities. Medical understandings of the relationship between alcohol and reproduction have varied through time, but diagnoses of alcohol's effect on offspring have always reflected broader social preoccupations with both social and biological reproduction and their cultural meanings. Medical ideas about alcohol and reproduction have tended to align with wider social norms, particularly in matters of gender. Thus, one striking continuity between past and present medical ideas about alcohol and reproduction is that, over the last two centuries, pregnancy has been used as a location to project social anxieties and to exert social control. The kinds of social anxieties and beliefs embedded in the contemporary clinical diagnosis of FAS bear striking parallels to past knowledge. As the Deborah Zimmerman case makes clear, the diagnosis of FAS is not merely a medical categorization; it contains a significant moral element as well. This confluence of the medical and the moral is one of the most significant aspects of the modern diagnosis of FAS, as well as one of its strongest links to past ideas about alcohol and reproduction. Social norms help to determine diagnoses, and diagnoses in turn can shape social norms. In the case of FAS, norms about female behavior, and specifically *maternal* behavior, helped to create the possibilities for the discovery of FAS and its codification as a clinical diagnosis. But that clinical diagnosis in turn has profoundly shaped norms of prenatal behavior in the United States today.

As Deborah Zimmerman's story suggests, constructions of FAS — like constructions of abortion[40] — also incorporate deep cultural anxiety over changes in gender roles, particularly notions of motherhood. What constitutes "good" motherhood in our society? This question became particularly difficult to answer in the last decades of the twentieth century, when women increasingly moved out of the home and into the paid workforce and when the breakdown of traditional family structures accelerated. Yet the appropriate role for women was also a matter of significant social dissent in the late nineteenth century, when women began to struggle for such rights as suffrage, divorce, and education, traditionally the exclusive prerogatives of men. In both eras, when the social order threatens to disintegrate, women are castigated for their failure to uphold the traditional roles that are seen as the foundation of that order.

In the past, doctors and laypeople alike believed in the doctrine of maternal impressions: the idea that something a woman experienced during pregnancy — a shock or sudden fright or a craving for a particular food — could leave a lasting "impression" on her unborn child. A woman who hungered for strawberries, for example, might give birth to a child with a "strawberry" birthmark. The sudden sight of a hare leaping across her path might cause a child to be born with a harelip. Many of the maternal impression myths are about female "selfishness" or about women desiring things *for themselves*.[41] In American culture, women who are not selfless, self-sacrificing, and self-abnegating transgress important boundaries. The ancient notion of maternal impressions continues to hold sway over not just the popular imagination but in many ways the medical imagination as well.[42] In the interviews I report on in Chapter 4, for example, one of the doctors said, "I personally think the true fact is whatever you [the pregnant woman] do does affect the baby." Thus, a woman's behavior during pregnancy reveals her moral character as well as her fitness to become a mother. To drink during pregnancy is to pollute one's womb; the infant with fetal alcohol syndrome is himself a piece of social pollution, undesirable, unwanted. In imagining the mother and the fetus at odds, we acknowledge a break in a previously ordered relationship. FAS is not just a cluster of physiological disorders, it also symbolizes a disordered role — a mother, like Deborah Zimmerman, who is not maternal.

Why is fetal alcohol syndrome as a kind of *prenatal* risk particularly interesting or informative? Pregnancy crystallizes concerns about gender, female identity, motherhood, and work, as well as hopes and fears for children — the next generation, the "future" of society. These abiding concerns about gender, female identity, motherhood, and work are all crystallized around reproduction — and pregnancy in particular. Even the briefest history of medical and lay ideas about pregnancy would reveal what a powerful prism pregnancy constitutes, refracting and reflecting notions of risk and responsibility, uncertainty, and social desires. Pregnancy *embodies* risk. Moreover, pregnancy offers insight into the relationship between one generation and the next, as well as the relationship between the human body and its environment, morality and the management of deviance, our fear of uncertainty and our need to control it. Pregnancy, which is "essentially uncontrollable,"[43] is particularly open to becoming a social metaphor. It is vulnerable to multiple and conflicting constructions because it is a kind of "black box" that may contain any number of meanings; its "true" meaning is invisible and protean. Many essential contradictions inhere in pregnancy. Becoming pregnant is both a quotidian human event, one that takes place thousands of times daily, and an awesome, mysterious force through which new life is created. It is both sacred and profane, and it can be both fragile and tenacious, precarious and persistent. Yet, paradoxically, in many ways pregnancy remains beyond human control: our mastery of the cryptic processes of conception and gestation is far from complete. As prone to human manipulation as the process of reproduction has become in our society, many people still regard conception, pregnancy, and birth as everyday miracles. Indeed, "pregnant women in American society are also marked through ritual practices and social beliefs as having a sacred character and as experiencing a sacred event."[44] Reproduction is still invested with the aura of the sacred.

Another reason that pregnancy is subject to sustained social attention is the duality inherent in it. For pregnancy involves not just the pregnant woman but her future child as well, not only this generation but the next, not only present but future. Although, at the most fundamental level, pregnancy is merely an individual physiological state, it nevertheless embodies multiple, overlapping sets of social relationships. Pregnancy, as we all know, is typically the result of an intimate interaction between a man and a woman, an interaction that is at once physical and social. Pregnancy is moreover the catalyst for one of the most significant social dyads, between a mother and child, and pregnancy is part of what makes families — the quintessential social unit — come into being. Pregnancy also involves the relationship between human generations. And, because reproduction is so often politicized, pregnancy involves the relationship between the individual woman and the society of which she is a part.

This duality in pregnancy comes to a head in the notion of maternal-fetal conflict, which emerged in part from the abortion debate in the United States. But the roots of this conflict are much deeper and wider, encompassing social conflict over profound changes in gender roles and swiftly evolving medical knowledge and technology. Prosecutions like that of Deborah Zimmerman represent the social enactment of the concept of maternal-fetal conflict; and they demonstrate that medical ideas can have profound effects on individual bodies and lives. Indeed, as our society has increasingly emphasized the individual over the communal, pregnancy has assumed new shades of meaning. For if pregnancy embodies risk, it also embodies responsibility. Was Deborah Zimmerman alone responsible for the outcome of her pregnancy, as the attempted murder charges against her seem to imply? Or do we as a society share responsibility for ensuring healthy births, as the doctors who intervened and persuaded her to consent to a cesarean section seemed to believe? As the notion of maternal-fetal conflict suggests, pregnancy today seems to lie at a crucial nexus between individual and social responsibility, illuminating ideas about risk and responsibility at both levels.

Because pregnancy happens only to women, constructions of it (not to mention the experience of it) are profoundly gendered: pregnancy affords an arena both to express and to enact ideal (and idealized) roles for women and to exert control over deviance. Thus, since fetal alcohol syndrome is a condition that occurs in pregnancy, it affords a special focus for moral opprobrium; the fetus is the "point of intersection of many social processes."[45] Moreover, pregnant women are construed as a kind of public property in American society.[46] They are subject to intense moral scrutiny, to monitoring, and to overt social control of their behavior. A woman's actions during pregnancy are regarded as revealing her moral fiber, as well as her potential fitness as a mother.

The irony is that even as women have achieved more gender equity and greater accomplishments in traditionally male domains, they have in some ways become ever more tightly identified with the biological function of reproduction. Even as the women's liberation movement broadened the categories of acceptable behavior for women, the prevailing ideology about pregnancy and the meaning of motherhood circumscribed women's roles within another realm. Historically, at moments of revolution in gender relations and roles, reproduction has become a tool to manipulate women — by defining them solely in terms of their reproductive function, by accusing them of failing society because they are producing too few children or children of inferior quality. At these moments, women are explicitly expected to bear responsibility for the social order, as well as for offspring. In the last quarter of the nineteenth century and again a hundred years later, at the end of the twentieth, women collectively strove to achieve major changes in their social status and roles. And in both eras, medical attention to reproduction — and specifically to women's actions during pregnancy — intensified.[47]

As the case of alcohol consumption shows, responsibility for reproduction has shifted over time, from something jointly shared by men and women to a duty that rests solely with women. Or, more accurately, a responsibility that resides *physically* and *primarily* in women's bodies. Women not only bear children, they bear the burden of ensuring the future of society. One consequence of this shift is increased social preoccupation with pregnancy — and desire to control women's actions, behaviors, and choices during pregnancy. An often-invoked metaphor to describe this phenomenon is that of the "pregnancy police." It is an apt term. Just as we rely on the actual police force to maintain social order, particularly in times of crisis, the pregnancy police are marshaled to monitor and restrict women's behaviors in the name of social as well as biological order. The social desire to control pregnant women stems from belief that pregnancy has consequences for society as well as for individuals. Drinking is one of the primary sites where such monitoring occurs.

Yet social investment in reproduction cannot alone explain the rising paradigm of maternal-fetal conflict. I have so far discussed the emergence of maternal-fetal conflict as a consequence of changes in medical technology and medical knowledge, changes that diffused to the lay public as well. But the invention of ultrasound, intrauterine photography, electronic fetal monitoring, fetal surgery, and like technologies cannot alone account for the profound transformation in medical and lay thinking about the fetus over the last thirty years. It is also the meaning of the fetus that has changed. The fetus has become more childlike. Antiabortion activists regularly and systematically invoke an imagery and language of "babies" to talk about the fetus. What we are witnessing is not only the ascendance of a paradigm of maternal-fetal conflict but also an extension of childhood as a social and biological state, an extension backward, into the womb. Certainly Joan Korb saw Megan Zimmerman as a *child* worthy of state protection even *before* she was born, since Deborah Zimmerman was never accused of attempting to kill her daughter *after* she was born. Likewise the hundreds of other prosecutors across the country who have invoked child abuse statutes to target women who use illicit substances during pregnancy. Although such language and action are often interpreted as an attempt on the part of the foes of abortion to create fetal personhood and thus delegitimize abortion through the back door, so to speak, a deeper cultural process is going on as well. The fetus is becoming more personlike, but not solely through the efforts of antiabortionists. Today it is not uncommon for pregnant women to leave their first ultrasound appointment at eighteen weeks with a grainy image of a fetal blur, labeled "baby's first picture." Not only is pregnancy sacred, but the fetus itself has been sacralized. Just as in the nineteenth century childhood was sacralized and made a separate and distinct phase of life, in the late twentieth century a similar process unfolded. Fetalhood is becoming a distinct and separate phase of childhood. The social consequences of this are profound, but especially so for women, whose bodies literally carry fetuses and cradle children at once. The diagnosis of fetal alcohol syndrome frames a window of exceptional clarity onto our society precisely because it encompasses these twin social phenomena of maternal-fetal conflict and sacralization of the fetus.

The history of medical understandings about alcohol and reproduction elucidated in this book implies that the need to classify physiological disorder stems as much from perceived social chaos as from any imperative to advance medical knowledge. The discovery of the "disorder" of fetal alcohol syndrome reflects not only a recognition of the pattern of growth retardation, neurological dysfunction, and craniofacial abnormalities that is the core of the diagnosis of FAS but also a need to impose order on a disorderly society. In 1973, the United States was a society fraught with race, class, and, in particular, gender tension as it weathered profound social upheavals reordering the relations between whites and blacks, the privileged and the disadvantaged, men and women. It was a society endangered by environmental degradation and fearful of toxins, from thalidomide to DDT, that seemed to threaten the very survival of the human race. It was a society captivated and repulsed by advances in technology, from nuclear power to fetal ultrasonography. It was a society increasingly driven by an ethos of individualism that elevated self-fulfillment to a paramount goal, a society that urged women both to step forward by burning their bras and shrugging off male chauvinism and to step backward by eschewing modern medicine and returning to a more authentic experience of "natural childbirth." Finally, as the still-reverberating U.S. Supreme Court decision in *Roe v. Wade* continues to demonstrate, it was a society deeply divided about the kinds and extent of reproductive freedoms to grant to its female citizens. Medicine seeks to create epistemological order by classifying and categorizing diagnoses; however, medical knowledge reflects and magnifies social orders as well. Fetal alcohol syndrome — a diagnosis applied to children who are quite literally disordered and disorderly — embodies the social conflicts of the modern era.

Medical knowledge shapes the choices we face as individuals and as a society. In the case of fetal alcohol syndrome, medical knowledge dictates how individual women should behave during pregnancy. However, even as it frames individual choices and responsibilities, the diagnosis of FAS elides public choices. Knowledge, once constructed, can be used to obscure certain dimensions of reality, even as it illuminates others. Drinking during pregnancy is stratified by race, education, and social class; risk is not randomly distributed through the population. Despite a public rhetoric of risk that avers all women are equally at risk, women who suffer social disadvantage are far more likely to bear the burden of adverse outcome. Thus, we witness in the example of fetal alcohol syndrome an inversion of the distinction C. Wright Mills observed at midcentury:[48] our prevailing understanding of this disorder turns "public issues of social structure" into "private troubles of milieu" — and even more significantly, into opprobrium that is targeted at individuals rather than at the social forces that shape their choices. Is the disorder we diagnose as fetal alcohol syndrome caused by exposure to alcohol in utero, or is it caused by poverty, social distress, and the deeply rooted inequality of American society? The rest of this book explores how the assignation of etiology and responsibility to health and illness reflects our sense of the proper social order no less than our understanding of particular biological processes.

CHAPTER 2

THE "QUESTION OF ALCOHOL AND OFFSPRING" IN THE NINETEENTH CENTURY

In discussing the question of alcohol in its relationship to heredity, it may be well at the outset to note the fact that people generally take a very definite and emphatic stand upon one side or the other of the question.

—C. C. Wholey, M.D.

Alcohol has been an integral part of most Western societies for centuries, and speculation about the relationship of alcohol to reproductive process and outcome has a long history. For millennia before embryology was well understood, observers tried to explain the mysterious process of human reproduction; often, speculation about alcohol's effect on offspring was part of this endeavor. Ideas about heredity were diffuse and varied, though they centered around the observation gleaned from agriculture, animal husbandry, and human experience that "like produces like." Thus, alcoholic parents were believed to produce alcoholic progeny. However, the effect of parents' consumption of alcohol was not limited to the creation of successive generations of alcoholics. It was also widely believed that drunkenness in the parents might result in both physical and mental defects in the child.

Although most of these early ideas about alcohol and reproduction bear faint resemblance to the modern conception of fetal alcohol syndrome, they nevertheless reflect striking parallels in the construction of medical knowledge and in its uses. In the nineteenth century, physicians accumulated a body of evidence that purported to link alcohol to heredity. Yet the certitude with which they assembled that knowledge and interpreted it to make arguments about social policy was disproportionate to the extent of the evidence, and different physicians often interpreted the same set of "facts" in ways that led to radically different medical and social conclusions. By examining closely the kinds of evidence and ideas brought to bear on the question of alcohol and offspring, we can see how, in the past as in the present, medical science and moral certitude combine to produce seeming certainty where in fact there is much uncertainty. Physicians did not hesitate to draw firm conclusions based on speculation and supposition. As they constructed a body of medical knowledge about the relationship between alcohol and reproduction, physicians expressed concerns about responsibility not only for individual birth outcomes but for the composition and future of society more generally. In fact, medical beliefs about the effect of drinking on offspring often expressed ideas about social relations and the social order. Physicians thus used the certainty forged through medical knowledge to help direct social policy in ways that prefigure our own social responses to drinking and reproduction.

Yet doctors in the eighteenth and nineteenth centuries conceptualized the relationship between alcohol and reproduction in ways very different from our current notions of fetal alcohol syndrome. Most significantly, in this period, physicians were concerned not with a prenatal effect of alcohol on children but with what they called a hereditary effect. Medical understanding of heredity in this pre-Mendelian period still drew deeply on ancient thought—particularly on the significance of the moment of conception and on the doctrine of maternal impressions. Classical thought postulated that the parents' mental, emotional, and physical state of being at the moment of conception left a lasting imprint on the child. Whether parents—and here the belief involved both the father and mother—were giddy or sullen, somnolent or elated, sober or drunk at the instant of conception determined their child's mental and physical character. Thus, parents who were inebriated during intercourse were believed to beget children who were weak, "crooked," or "wont to be fond of wine."

Classical texts contain many allusions to this belief. A variety of ancient texts speculate about the effect of alcohol on libido and sexual performance, and several contain explicit references to the effect alcohol exerted on offspring. A passage in Plato's Laws, for example, contends that "it is not right that procreation should be the work of bodies dissolved by excess of wine, but rather that the embryo should be compacted firmly, steadily and quiet in the womb.

But the man that is steeped in wine moves and is moved himself in every way, writhing both in body and soul; consequently, when drunk, a man is clumsy and bad at sowing seed, and is thus likely to beget unstable and untrusty off-spring, crooked in form and character."[1] Likewise, Plutarch writes in *Education of Children* that "children whose fathers have chanced to beget them in drunkenness are wont to be fond of wine, and to be given to excessive drinking."[2] As these examples show, the focus in the ancient world was primarily on the man's role in procreation, reflecting the belief that the woman was merely the receptacle for the man's generative seed. However, Soranus of Ephesus, the so-called father of gynecology, also considered the impact of the woman's sobriety on the child. Writing in the second century a.d., he stated, "In order that the offspring may not be rendered misshapen, women must be sober during coitus because in drunkenness the soul becomes the victim of strange phantasies; this furthermore, because the offspring bears some resemblance to the mother as well not only in body but in soul."[3]

The idea that the child resembled the mother "not only in body but in soul" was the foundation of the doctrine of maternal impressions, which posited that the child bore the physical imprint of the mother's ineffable soul, particularly feelings of fright or longing. The developing fetus was therefore vulnerable to the mother's moods and whims. The doctrine of maternal impressions was in fact the major explanation for anomalous births, including malformed children and infants with a harelip, clubfoot, cleft palate, or birthmark. Monstrous births — both real and imagined — were explained under this rubric. The woman's "ardent and obstinate imagination"[4] might result in the birth of a child with a harelip after she saw a rabbit, for example, or the birth of a child with a raised, red birthmark if she felt a hunger for berries that went unfulfilled.[5] Significantly, the doctrine of maternal impressions rests on the notion that a woman's cravings, desires, and wants could "mark" the child she carried in highly visible and symbolic ways. Often such births were interpreted as omens or portents, particularly of a disturbance in the social or natural order. In the sixteenth century, Ambroise Paré, chief surgeon to Charles IX and Henri III of France, contended that such births "are usually signs of some forthcoming misfortune."[6] Thus, birth defects have long been viewed as disturbing physiological manifestations of disruptions to the social order.

Although these premodern texts demonstrate a concern for the effect of alcohol on offspring, they do not imply any cognition of a *fetal* effect. Rather, they suggest that drunkenness during conception, like any other mental disturbance at this crucial moment, could cause the child to be ill-formed.[7] Indeed, neither the belief in the importance of parental emotions at the moment of conception nor the doctrine of maternal impressions focused particularly on the role of *alcohol* in reproduction. Both involved much wider speculation on the multiple causes of disordered births.

Alcohol and Social Disorder: The Gin Epidemic

In the early modern world, concerted attention to the influence of parental drinking on children was initially provoked by England's Gin Epidemic in the first half of the eighteenth century. In order to provide new markets for grain and to reduce imports, the English government lifted traditional restrictions on distilling alcohol in 1720. The country was flooded with cheap gin, and the drinking of spirits, formerly reserved for the upper classes, became a luxury even the poorest could afford. Indeed, to many observers, gin drinking seemed especially prevalent among the poor, for whom gin served as an all-purpose analgesic against the great miseries of life, including cold, hunger, and despondency. Furthermore, distilled spirits were widely used medicinally.[8] National consumption of gin rose from 2 million gallons in 1714 to 5.5 million gallons in 1735 and 11 million gallons in 1750.[9] It was not long before the social effects of such prodigious consumption were felt. Although most commentators were concerned primarily with crime, vice, and general social disorder among the lower classes, they often expressed dismay over increased fetal and infant mortality as well. By 1725, the College of Physicians petitioned Parliament to tax gin, claiming that gin was "a cause of weak, feeble and distempered children."[10] A 1736 report to Parliament noted, "Unhappy mothers habituate themselves to these distilled liquors, whose children are born weak and sickly, and often look shrivel'd and old as though they had numbered many years."[11] Henry Fielding, campaigning against the increase in crime precipitated by cheap gin, queried, "What must become an infant who is conceived in *Gin*? with the poisonous Distillations of which it is nourished, both in the womb and at the Breast."[12] The specific effect of alcohol on infants and children thus served as an important social alarm and helped to provoke legislative action in response to the perceived crisis. England's Gin Epidemic ended with passage of the Act of 1751, which raised taxes on distilled spirits and curbed their retail sale.

Although the perceived rampant social disorder of the Gin Epidemic dissipated in the latter half of the eighteenth century, the greater consciousness of the disruptive role of alcohol that the Gin Epidemic had stimulated — disruption in society generally and in procreation specifically — persisted into the nineteenth century. In both Britain and the United States, views shifted towards thinking of alcohol as a potent source of social evil, and in the United States an organized temperance movement began to emerge as early as the first decade of the nineteenth century. Concurrently, physicians began to note more systematically the relationship between alcohol and offspring. The evils of the hereditary effects of alcohol added another dimension to general social concern about alcohol consumption. This focus on the future burden to society of children who are the products of alcoholic conceptions is a theme that echoes from the time of the Gin Epidemic until the present.

Alcohol and Heredity

By the turn of the nineteenth century, alcohol was increasingly targeted as a source of social evil. During this period, medical ideas about alcohol and reproduction rested on broader notions of the role of heredity as a determinant of both individual and social destiny. Two broader social currents — concern about alcohol and concern about heredity, specifically the fear of degeneration — flowed together, thereby generating a growing torrent of attention to and concern about alcohol's effect on children.

First of all, the problem of inebriety was a major social rubric of the time. Drunkenness, or the "drink crave," was used to explain a host of social ills, including crime, illegitimacy, and unemployment. A formal temperance movement began in the United States in the early 1800s, though it did not achieve widespread social salience until sometime after the Civil War. By the late nineteenth century, the medical temperance movement had spawned its own journals, including the *Journal of Inebriety*, edited by the Philadelphia physician Thomas Davison Crothers and first published in 1871 by the American Medical Association for the Study of Inebriety and Narcotics, and the *Quarterly Journal of Inebriety*, first published in 1876. The articles and letters in these journals increasingly focused on the debate about alcohol and heredity, and by the 1870s the notion was widely accepted among physicians and laypeople alike that alcohol was injurious to offspring.[13] Heredity and inebriety were thus two major social preoccupations of the nineteenth century, and they became linked in two very different social movements: temperance and eugenics. Although inebriety was a major focus of medical investigation and advocacy in the nineteenth century, the early temperance movement turned its attention only occasionally to the effects of alcohol on offspring. "Children born of intemperate parents bear in the birth the germs of disease, die prematurely, or drag along a languishing existence, useless to society, depraved and possessed with evil instincts," noted an American physician in 1806, echoing observations made during the Gin Epidemic.[14] His opinion was the exception in the first half of the century, but his words contain the roots of later concerns that would link alcohol and offspring — offspring "useless to society" — and the social order.

The concern about alcohol's hereditary effect had two dimensions. Physicians believed that inebriety or drunkenness could be passed directly from parent to child; thus, the children of parents who were drinkers were more likely to be drinkers themselves. Doctors also believed that drinking in the parents led to other kinds of constitutional weakness in their children — feeblemindedness, insanity, even such diseases as scrofula and tuberculosis. Both of these beliefs reflect the core assumption that "like begets like," the foundation for nineteenth-century understandings of heredity. Certainly these ideas about heredity were far removed from our present understanding of genetics. Most important, for most of the nineteenth century there was little attention to or theory about the precise biological pathways through which heredity was manifested. Much older beliefs about the parents' state of mind at the moment of conception and about maternal impressions persisted, providing at least a vague if inchoate explanation for the mechanism of heredity for most of the nineteenth century.

Second, most doctors, scientists, and laypeople in this period assumed the truth of the doctrine of the heritability of acquired characteristics, a doctrine most often identified with the French biologist Jean-Baptiste Lamarck. Parents' cumulative life experiences could have a formative effect on their children. Parents who were exhausted from overwork, for example, were liable to produce weak and sickly children, who literally inherited their parents' fatigue. Parents who took up the drink habit could pass this evil on to their children through some direct if unspecified line of heredity.

These beliefs were crystallized in the notion of diathesis, or constitution, an idea dominant among physicians and laypeople throughout the nineteenth century and well into the twentieth century. According to this idea, individuals are endowed with a vigorous or weak constitution, protecting them from or predisposing them to particular illnesses.[15] In the nineteenth century, the term *diathesis* was used more specifically to refer to particular constitutional types such as tuberculous, bilious, cancerous, gouty, and nervous.[16] The term itself has a long history, dating back to Aristotle; it enjoyed a renaissance beginning in about 1800.[17] A diathesis was a predisposition to a chronic condition that might appear, recede, and reappear throughout a person's lifetime — "namely a high susceptibility to a disease of its particular kind."[18] Moreover, a diathesis could be inherited. "It is generally observed that the constitution, the temperament and diathesis of the offspring closely resemble the parent; and that whatever disposition to disorder, whether of structure or function, the latter may have possessed, it is liable to evince itself in the former."[19] Hereditary diathesis thus accounted for both differing susceptibilities to disease among individuals and variation in rates of affliction among families.[20] Although alcoholism was not considered the consequence of a specific diathesis, as was tuberculosis or scrofula for example, it was commonly held that alcoholism, like pauperism and criminalism, could be passed down in families from one generation to the next.[21] Thus, the basic ideas undergirding diathesis — that people demonstrate lifelong proclivities and that these proclivities may be inherited — were also the foundation for thinking about alcohol and heredity. Moreover, the progeny of the drinker inherited not just a tendency to drink but a package of constitutional weaknesses. As one physician explained, "The offspring of the habitual drunkard generally inherits such degenerative conditions as idiocy, scrofula, deaf-mutism, the tendency to phthisis and sometimes epilepsy, while that of the dipsomaniac is liable to the more active or spasmodic forms of nervous disease, such as suicide, acute mania, epilepsy, and crime."[22]

By midcentury, new methods of social science investigation — descriptive statistics and the principle of statistical inference — allowed physicians to confirm the relationship they had hypothesized between alcohol consumption in parents and constitutional weakness in their progeny. In order to prove their assumptions, observers shifted their focus from individuals to populations. Aggregate social statistics — essentially, counts of people or of vital events like birth and death — were being more widely employed in the early 1800s, in England in particular. Physicians interested in temperance exploited two different ways of viewing the relationship between alcohol and offspring, using these new social statistics. In the first, they looked at populations of "defectives" — the institutionalized insane or "feebleminded," incarcerated criminals, or paupers — to find evidence of drunkenness and inebriety among their parents. In the second, they looked among parents who drank for evidence that their children suffered as a result.

In England, an 1834 report to the House of Commons from the Select Committee on Inquiry into Drunkenness was one of the first systematic social inquiries into the effect of alcohol on children. The report cited the testimony of Dr. R. G. Dods that the children of intemperate parents are less healthy than those of temperate parents, "as intemperate parents . . . give a taint to their off-spring, even before its birth."[23] Although the word *taint* recalls the notion of maternal impressions, the focus here is firmly on both parents. Dr. Dods testi-fied that the "infant [of alcoholic parents] has frequently not merely the want of a healthy aspect, of plump, round outline, but a starved, shrivelled and im-perfect look."[24] Here, particularly in the description of these babies as shriv-eled, the 1834 report echoes similar testimony to Parliament almost a hundred years earlier, during the Gin Epidemic.

In the United States, many state governments undertook similar investi-gations of the effect of alcohol on offspring, particularly alcohol's contribution to so-called defective populations: the idiotic, imbecilic, insane, and otherwise institutionalized. Samuel G. Howe's 1848 report to the government of Massa-chusetts found intemperance to be a leading cause of deterioration in families and cited alcohol as "a prolific cause of idiocy."[25] Of the 359 idiots studied by Howe, 99 were found to be the children of drunkards. Howe noted that the children of the intemperate are "very apt to be feeble in body and weak in mind" and that "idiots, fools and simpletons" are common among the progeny of drinkers. An 1856 report to the General Assembly of Connecticut related similar findings, attributing 76 of 253 cases of idiocy to intemperance in the par-ents.[26] In 1877, Nathan Allen, who bore the marvelous title of Commissioner in Lunacy to the Commonwealth of Massachusetts, confirmed Howe's earlier findings and reported that three-quarters of the inmates of the state's criminal, reformatory, and pauper institutions were children of inebriates.[27] These re-ports contributed to the growing perception that alcohol was responsible for creating a large social burden.

As these reports indicate, the growing concern about alcohol and repro-duction first found expression in studies of institutionalized populations, such as the inmates of prisons and homes for imbeciles and insane asylums; how-ever, by the middle of the nineteenth century, investigators cast their nets more widely to the population at large.[28] No longer did they look for the effect of al-cohol only in certain captive populations where they knew defectives were concentrated, but they sought evidence for alcohol's effect among the general population. Physicians now turned their attention to defective offspring in "alcoholized" populations.

One kind of population study examined the effect of alcohol among groups that had special exposure to it, such as "calendar studies" of the timing of births of defective children correlated with drinking season. In Switzerland, Bezzola conducted "a statistical investigation into the role of alcohol in the origin of innate imbecility."[29] He studied the correlation, for the years 1880 to 1890, between the annual periods of greatest drinking (around the New Year and from April to June, the period of Easter and weddings) and the incidence of imbecility in children born nine months later, and he contrasted this with the incidence in children born nine months after July and September, "the months of unusual labor," when there was much less drinking. He found that the rate of birth of imbeciles was higher than expected for children conceived from January to February and from April to June, and it was dramatically lower for children conceived between July and October.[30] As he vividly described the effect, "The influence of Bacchus, Gambrinus, and the Schnapsdevil combine with that of Venus to send the pathological curve to its highest elevation"[31] — the height of the curve demonstrating the powerful hereditary effect of alcohol. In seeking to correlate the season of conception with birth outcome, not only did Bezzola draw on the ancient notion that the parents' intoxication at the time of conception had a permanent effect on their offspring, he also sought to quantify the effect.

Bezzola's crude natural experiment was paralleled by others that used region or time to enumerate defective offspring where or when alcohol consumption was particularly high. One such study noted a 150 percent increase in the incidence of "congenital idiocy" in Norway following removal in 1825 of the spirit duty.[32] The *Quarterly Journal of Inebriety* reported that in Fauborn, France — near Limbourg-sur-Lahn, an area with a heavy concentration of distilleries — "the population appears to be composed of cretins and examples of degeneration of all descriptions."[33] Although many of these studies were carried out in Europe, they were widely reported in the American temperance literature.

As such studies suggest, the ancient notion that drunkenness at conception affected offspring persisted through the nineteenth century among physicians and laypeople alike. The idea was succinctly stated by the Philadelphia physician and vehement prohibitionist Thomas Davison Crothers: the "physical health of the parents at the time of conception controls the future of the child to a large degree."[34] However, the belief encompassed more than the physical well-being of parents, extending to their mental and emotional states; even the father's occupation could have a definitive influence on offspring. Throughout the nineteenth century, in fact, physicians and laypeople believed that intoxication at the time of conception was particularly dangerous but that any alcoholism in the parents predisposed their progeny to a spectrum of constitutional weaknesses. This belief often took what in retrospect seem to be extreme forms, in the sense that the parents' experience was seen as a direct template for the child's life. Physicians and educated laypeople generally accepted the notion that acquired characteristics, including patterns of behavior, could be inherited. Dr. Crothers collected case studies showing the effect of parents' alcoholism on their children. Like other physicians of the day, Crothers invested remarkable explanatory power in the concept of heredity. He reported some of his cases in Philadelphia's *Medical and Surgical Reporter* in 1887. In these cases, life experience itself is seen as heritable, as parents' drinking histories are mirrored exactly in their children. In one illustrative case, the son of a sea captain was temperate for long stretches at a time, like his father — who drank only when ashore after one or two years at sea, and then only in seclusion with his wife. The son replayed this pattern — suddenly he would run off, procure a female companion and a room, and, in the woman's company, drink to excess. Crothers postulated that "how the periodical drink manias of the parents should so impress the child as to be reproduced, antagonized by the circumstances and surroundings of life, is a mystery [that] . . . can only be answered by a study of heredity" (550). In another case, Crothers described a patient who was "temperate up to forty years of age, when suddenly he drank to intoxication, and was seen on the streets shouting in a delirious way" (550). In diagnosing this patient, Crothers noted that the man's father was a temperate man who married for the first time after forty and "soon after drank to great excess." The son was conceived in one of these "drink paroxysms." Crothers believed that "the remarkable similarity of the symptoms of father and son, coming on at the same time of life . . . seem to indicate a direct heredity from father to son" (551). Crothers's conclusion was shaped not only by his belief in the heritability of acquired characteristics but by the older notion that the parents' state at the time of conception was profoundly influential on the child's constitution.

Crothers's case studies also suggest that nineteenth-century observers believed in "the transmissibility of the 'drink crave,'"[35] or the predisposition of the children of intemperate parents to become drunkards themselves. Children of alcoholic parents were "predisposed by their very organization to have cravings for alcoholic stimulants."[36] The 1834 Report to the House of Commons had also noted that the children of drinkers have "a hereditary predisposition to the use of ardent spirits."[37] However, the dangerous inheritance that alcoholic parents passed on to offspring went far beyond the drink crave. The hereditary effect of alcohol was thus twofold: inebriate parents endowed their children with both a weakened constitution and a predisposition to drink, setting up a vicious cycle. Some observers tried to locate the origins of that cycle, yet whether alcoholism weakened constitutional stock or weak stock predisposed one to alcoholism was a question not easily settled. Doctors argued that "if feeble-mindedness is not due to alcoholism, on the other hand alcoholism is to a very large extent due to feeble-mindedness."[38] Doctors were thus preoccupied with the effect of alcohol not only on individuals, not only on the present generation, but on society and on future generations as well.

Physicians also used family lineages, such as those depicted in Figure 1, to illustrate the notion of a persistent, hereditary, constitutional bent towards alcoholism and worse. Thorough investigations of family lineages were among the most important types of evidence that physicians mustered to verify the hereditary evils of alcohol and later to justify eugenic policies. Physicians reported on patients under their care, constructing elaborate pedigrees in which the peculiar manifestation of degeneration was noted in each of the ancestors and each of their descendants. The Jukes, depicted by Richard Dugdale in 1877, and the Kallikaks,[39] described by Henry Goddard in 1912, were the most famous of the alcoholic lineages in the United States. Typical of these alcoholic lineages is one reported in a Detroit medical journal in 1902:

> Frau Ada Jurke, who was born in 1740, and was a drunkard, a thief, and a tramp for the last forty years of her life, which ended in 1800. Her descendants numbered 834, of whom 709 were traced in local records from youth to death: 106 of the 709 were born out of wedlock; there were 144 beggars, and 62 more who lived from charity. Of the women, 181 led disreputable lives. There were in the family 76 convicts, seven of whom were sentenced for murder.[40]

The impulse to quantify the ill effects of alcohol was a significant one. By presenting the precise numbers of descendants who possessed a hereditary taint

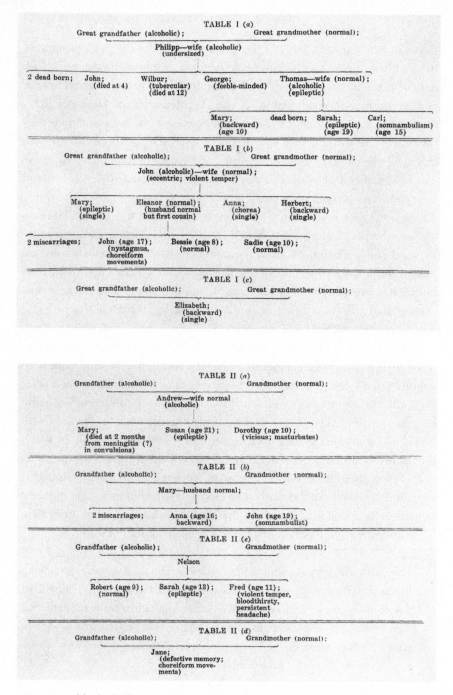

FIGURE 1. Alcoholic lineages illustrating degeneration. From Alfred Gordon, "The influence of alcohol on the progeniture," *Interstate Medical Journal* 23, no. 6 (1916).

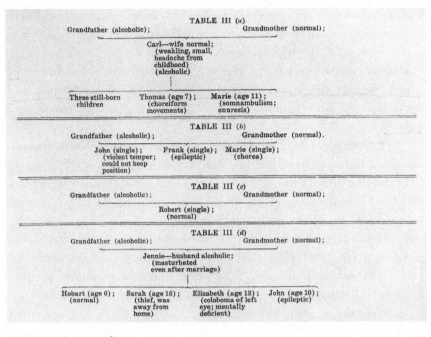

FIGURE I *(continued)*

from alcohol, doctors hoped to dramatize the extent of the problem as well as its urgency. Numbers lent precision — and an aura of certainty — to their efforts. Moreover, quantification was taking on greater importance in medicine and in social thought more generally. Alcoholic lineages enabled physicians to play the counting game and to muster a new kind of evidence for the hereditary effect of alcohol; the lineages dramatically magnified the kind of evidence that Crothers's case studies presented.

These lineages vividly illustrate the evil hereditary effects of alcohol. As Alfred Gordon writes, "In the three generations, we find individuals with various degrees of mental deficiency, epilepsy, choreiform movements, tremors, eccentricity and violent temper; we find also miscarriages and deadborn children."[41] Alcohol, "the only pernicious agent" in these lineages, causes a broad swath of ill health, unfitness, and social maladaptation.

The alcoholic lineages also enabled doctors to illustrate another major theory on the connection between alcohol and heredity, that of degeneration. Popularized by the French psychiatrist Bénédict Auguste Morel, who published *Traité des dégénérescences physiques, intellectuelles, et morales* in 1857, the concept of degeneration held that individuals inherited not specific ills but a "neuropathic constitution which tended to deteriorate in a specific and progressive pattern from generation to generation."[42] Thus, Morel's theory extended the popular concept of diathesis and added a dynamic element to it. In Morel's thought, the process of degeneration involved physical, mental, and social decline and was cumulative, with each generation weaker than the one before.[43]

In the first generation, the degenerative tendency manifested itself as immorality, alcoholic excess, and brutal degradation. Hereditary drunkenness, maniacal attacks, and general paralysis were evident in the second generation. The third generation was characterized by a return of sobriety, but now accompanied by hypochondria, systematic mania, and homicidal tendencies. Finally, the fourth generation yielded progeny marked by feeble intelligence, stupidity, an attack of mania at age sixteen, then transition to complete idiocy and probable extinction of the family.[44] "Whatever, therefore, may be the peculiar feature of this transmission, one thing is certain — the whole tendency is *downward*, physically, mentally, and morally, not only by injuring the constitution itself, but by increasing the power and influence of the animal propensities at the expense of reason, the conscience and the will."[45]

Morel saw degeneration as "the long-term product of the formidable influence of hereditary predisposition."[46] The parents' lives could leave a lasting imprint on the child. The concept of degeneration was enormously influential, captivating both the public imagination and the minds of physicians; it held sway for almost a century. Not until the 1950s, in fact, did the term drop entirely from medical usage.[47]

Morel considered the development of chronic alcoholism to be typical of the process of degeneration.[48] Chronic alcoholism in the parents led to idiocy in the children, according to Morel. The hereditary constitution of the stock had been damaged by what he called the "intoxicating principle," in this case, alcohol. "There is no other disease in which hereditary influences are so fatally characteristic," said Morel; degeneration caused a "progressive decline in both adaptation and fertility."[49] Following Lamarck, Morel saw the alcoholic as a regressive example of the human species.[50] This theme of social disintegration became increasingly strident in the last decades of the nineteenth century and the first decades of the twentieth.

Although initial medical concern about drinking and reproduction centered on the effect first on the drinker then on the drinker's children, concern — and moral indignation — gradually widened to include the drinker's effect on society. Indeed, the seeds of this kind of concern were planted during the Gin Epidemic, when a large measure of the outrage at drinking among the lower classes was directed at women and the effect of their gin consumption on their children — which in turn reflected the fitness of society overall. The fear was of degeneration not only among individuals or families but within society as well. The 1834 report to the House of Commons, for example, maintained that alcohol weakened each succeeding generation, eventually threatening to bring about "the diminution of the physical power and longevity of a large portion of the British population." "By such habits [as inebriety]," the report stated, "one generation after another becomes more and more effeminate, till they scarcely deserve the name of human beings."[51] In the United States, Lyman Beecher, a Massachusetts clergyman and antialcohol crusader, struck a similar note: "The free and universal use of intoxicating liquors for a few centuries cannot fail to bring down our race from the majestic, athletic forms of our Fathers, to the similitude of a despicable and puny race of men."[52]

As Crothers's case studies suggest, a child's biological inheritance from his parents was more than a reflection of their emotional status at the moment of conception. It was really the cumulation of the parents' life experiences — the sum total of their sins of omission and commission over their lifetimes. As one doctor claimed, "there cannot be any reasonable doubt that the sins of the fathers (or mothers) are here most fearfully visited on the children."[53] The theory of degeneration formalized the conviction that "it is positively certain that by the fixed laws of hereditary descent the iniquities of the parents are visited upon the children unto the second, third and fourth generations."[54] Hereditarian writing frequently invoked this biblical image of children condemned to suffer for the sins of their fathers and mothers. Ironically, however, nineteenth-century notions of heredity also reflected a deep-seated faith in the malleability of offspring. Even as hereditarianism insisted that "the evil men do lives after them,"[55] it also placed control over children's destiny in the hands of their parents, whose behavior throughout life, not just prenatally, was a powerful determinant of their progeny's fate. The child "stands forth a complete composite picture of his progenitors."[56] Thus, notions of heredity revealed a remarkably labile social ideology that expressed both a powerful social stasis — children as predetermined replicas of their parents' failings — and an exuberant optimism about the ability of the present generation to protect and improve the lot of the next. Parents have "in a great measure, the character and destiny of [their] offspring, either for 'weal or woe,' under [their] control."[57] Samuel Howe had claimed in 1848 that parents could be "the blessing or the curse of the children; the givers of strength, and vigour and beauty, or the dispensers of debility, and disease, and deformity."[58] There was a wide degree of latitude in determining the fate of the progeny.

This dual optimism and pessimism about the role of alcohol in shaping human fate and about parents' ability to change their children's futures through sheer force of will was facilitated by the fuzziness of prevailing theories about the order of cause and effect in the association between alcohol and "weak stock." The persistent debate about the intertwined etiology of alcoholism and degeneracy muddied many of the theories advanced by nineteenth- and early-twentieth-century writers who often hesitated to determine definitively which condition caused, or at least preceded, the other: was alcoholism the result of some preexisting degeneracy, or did alcoholism cause degeneracy? Most thinkers, including Morel, equivocated, seeing alcoholism as both a cause and a consequence of degeneration. The British physician William Sullivan, for example, wrote, "Parental drunkenness . . . may get the credit of determining in the stock a degenerative tendency which really existed prior to it, and of which, in fact, it was merely a symptom."[59] Leonard Darwin (the son of Charles Darwin) noted that the "alcoholism of the parent and the defects of the children may both be the result of some common factor, such as inherited natural weakness of mind or body."[60] Indeed, such theoretical haziness best suited the competing ideologies of the Victorian period. By blurring the causal direction between inebriety and degeneracy, writers could express both dismay at individual decline and hope in the power of ameliorative policy to reverse social decline. Even as they decried degeneracy, they could identify its antecedents as well as opportunities to reverse the trend. After all, alcoholism could be eliminated through public health efforts and environmental reform — and moral effort on the part of drinkers. And eliminating inebriety alone, rather than attacking deeper social inequities, held the promise of curing not just individual drunkards but also society of the myriad ill effects of drunkenness. "Drunkenness is one of the greatest — perhaps the greatest — agent of degeneration at work among the human race, and to it must be attributed much of the disease, crime, moral obliquity and general degeneracy, physical, mental, and moral, which we find so common among the poorer classes in all large centers of civilization," proclaimed one social observer.[61] Thus, in individual patterns of alcohol consumption nineteenth-century reformers saw both the cause of and the solution to a wide range of social problems.

Despair about modern life was thus balanced by the sense that human agency could be brought to bear to cure these self-induced ills. The belief in the heritability of alcoholism in some sense reflected an inherent optimism in the world view of those who adopted it. Despite their apparent pessimism, Morel's theory of degeneration and American notions of hereditarianism were in some ways essentially optimistic: "since acquired characteristics could be inherited, environmental reform was all the more important."[62] Such optimism and faith in the ability of human beings to shape the nature of society is expressed in an 1890 article by Crothers in the *Journal of the American Medical Association*: "Recent studies of alcohol cases show that over seventy percent are directly inherited. If this is confirmed by later studies, the treatment of inebriety will in the future begin in infancy, and the higher science and art of medicine will win its greatest triumphs along the line of prevention."[63] This passage expresses a perfectionist sensibility, a belief that virtue as well as vice could be inherited, coupled with a faith in the potential dominance of nurture over nature. Perfectionists "saw in ameliorative policies and social engineering the route to the enhancement of the human constitution."[64] It was such a faith that lit the fervor of prohibitionists. One temperance doctor, for example, wrote, "There is much ground for hope that two or three generations, free from the social blight of widespread alcoholic indulgence, would largely free itself from the fearful consequences of alcohol."[65] The irony of this brand of American social optimism, particularly as expressed through the intersection of heredity, temperance, and eugenics, is that it spurred social reform efforts even as it located the source of social problems in individual morality and biology rather than in social structure.

Alcohol as Germ Poison

The quest to explain the observed world, to impose the reassuring order of analysis on the chaos of experience, intensified in the final decades of the nineteenth century. Early hereditarian thinkers did not trouble themselves with the mechanism of the hereditary effect, attributing it to vague pathways such as the parents' emotions or simply assuming that it existed. However, by the latter half of the century there was increasing interest in understanding how the hereditary effects of alcohol and other evil influences were transmitted. Although slow to take hold, August Weismann's theory of "continuity of the germ plasm"[66] (first published in Germany in 1886 and in England in 1893) introduced a new dimension into the way physicians and laypeople thought about the relationship between alcohol and offspring; they could now envision an etiological mechanism. Weismann's theory postulated that the germ cells — by which he meant the sperm or ova — of all living creatures contain "something essential for the species, something which must be carefully preserved and passed on from one generation to another." What Weismann and his contemporaries called "germ plasm" we today recognize as chromosomes, genes, and DNA. Although he lacked any knowledge of genetics and was not familiar with Gregor Mendel's experiments with peas, Weismann was able to formulate what remains essentially the foundation of modern genetics. In the absence of an experimental or biochemical understanding of the nature of germ plasm, Weismann and others invested the substance with semi-mystical properties. Germ plasm was seen as the essence of not only individuals but races. It was a kind of sacred trust to be guarded and protected from harmful influences.

Alcohol was one such harmful influence. By the last decade of the nineteenth century, scientists and social thinkers drawing on Weismann's theory had begun to speculate about alcohol as a "germ poison." This hypothesis resonated with the ancient belief that noxious influences at the time of conception adversely affected the offspring. The offspring was vulnerable at certain critical moments; the sins of the parents were recorded and coded into their germ plasm, to leave a lasting imprint on their progeny. Indeed, Weismann's theories had been prefigured in earlier thinking about heredity. His concept of the germ plasm provided the hypothetical medium through which the poisonous alcohol could work its degenerative effects. Thus, Weismann's theory did not so much revolutionize thinking as provide a formal explanation for relationships doctors already believed to exist.

This metaphor of germ poison also drew on notions of alcohol as bodily toxin, a "poison — the underlying evil of human life — the Mephistopheles of society."[67] Physicians conceived of intoxication as a type of "poisoning," explaining alcohol's somatic effect through its perceived toxic properties. According to Crothers, "The alcoholic . . . is literally a toxemic from poisons introduced into the body from without and poisons formed by chemical combination within, producing most complex disturbances and degenerations."[68] Thus, it was only natural that alcohol could be seen not only as a somatic poison but as a germ poison as well. Through the concept of the germ poison, the metaphor of poison proved a powerful description of alcohol's effect on both the individual and society. A physician warned in 1911, "This wholesale poisoning of civilized and semi-civilized races, if continued, will ultimately result in their extermination."[69]

Such notions were crystallized in Swiss psychiatrist Auguste Henri Forel's theory of blastophthoria, or germ deterioration. Forel advanced the notion that alcohol damaged the germ cells (that is, the egg and sperm) of women and men, leading to defects in their children.[70] Forel's theory had been foreshadowed by earlier thinking about the relation between alcohol and progeny, such as that of the Scottish physician Thomas Trotter, an early temperance advocate, who had argued in 1813 that "the organs of generation must equally suffer in both sexes, from frequent intoxication."[71] Indeed, the notion of alcohol as a germ poison grew at least partly out of the old idea that the parents' state of being at the time of conception was critical and that intoxicated parents conceived damaged children. The theory of blastophthoria provided both a hypothetical mechanism by which to explain this damage and a physiological pathway through which the process of degeneration might unfold. Moreover, Forel's theory of germ deterioration linked older notions of alcohol's damaging effect to the newer theories of the germ plasm as the mechanism for hereditary transmission.

According to Forel, "Alcohol, and probably also all other narcotics, poisons not only the individual, but also his sperm, the germs of his descendants."[72] "It is already well-known," he continued, "that inebriety, the chronic poisoning by alcohol, is transferred by the degeneration of the semen of man and of the ovaries of woman to the progeny of the drunkard. It is true, it very frequently causes in the descendants of the above-mentioned irresistible longing for alcohol, but also a variety of other diseases, the children of drunkards perishing in large percentage of debility of life, dwarfed growth, idiocy, mental diseases, and so forth."[73] Forel's theory of blastophthoria thus elaborated on Morel's theory of degeneration, postulating the exact somatic link between parents' sins and progeny's defects — the germ plasm itself. The notion of a germ poison resonated nicely with Morel's theory of degeneration, since it suggested that the defect would indeed be transmitted not just from parent to child but across generations.

The advent of new, more specific theories about alcohol's effect paralleled the development of new methods of scientific inquiry. In fact, Forel's theory pointed to the intersection of alcohol and eugenics. Just as, earlier in the century, medical concern about drinking had intersected with the social movement of temperance, in the final decades medical ideas about alcohol became entangled in the eugenics movement.

Alcohol and Eugenics

It was the hypothesized action of alcohol as a germ poison that first aroused eugenicists' interest in the relationship between alcohol and offspring. The term *eugenics* was coined in 1883 by Sir Francis Galton, a first cousin of Charles Darwin, and a scientist devoted to improving the human species through selective mating of marriage partners with superior genetic endowments. Although Galton was not a physician, many medical doctors wholeheartedly supported the eugenics movement, seeing in it the potential to cure both individual ills and social disease as well. Galton spent the early part of his career studying the inheritance of intelligence and came to believe that mental characteristics were as heritable in human beings as were physical traits. This conviction led him to advocate that "it would be quite practical to produce a highly gifted race of men by judicious marriages during several successive generations."[74] Other eugenicists believed alcohol could provide a short cut to this improvement of the human population through its property as a germ poison. However, the eugenics debate led doctors right back into the quagmire: did alcohol weaken a person's constitution, or did a weak constitution predispose one to drunkenness?

The debate about the causative order and relationship between intemperance and degeneration turned sharply divisive in 1910, when a pair of researchers in London challenged the prevailing view through the use of the new science of statistics: they claimed definitively that it was not alcoholism that led to bad stock, but bad stock that led to alcoholism. Karl Pearson was the founder of the Francis Galton Laboratory for National Eugenics, named in honor of his mentor. Ethel Elderton was one of Pearson's protegées.[75] They were statisticians, not physicians. And they took a decidedly more rigorous approach to investigating the effects of alcohol than had been undertaken in the past, no matter how meticulously earlier investigators had counted the children of drunkards among the insane, idiotic, epileptic, and other degenerates. Elderton and Pearson's first monograph was published in 1910, with the straightforwardly descriptive title A First Study of the Influence of Parental Alcoholism on the Physique and Ability of the Offspring.[76] The monograph was one of a series produced by Pearson that aimed to show the inheritance of intelligence and ability. Pearson, like Galton, was a devoted eugenicist and sought to use the new science of biometry, the application of statistics to biology and medicine — which he had invented with Walter F. R. Weldon — to demonstrate the truth of his eugenic convictions. In fact, Pearson and Elderton were not interested in the problem of inebriety or temperance; rather, their interest lay in what they saw as a much broader and more significant issue. They saw alcoholism not as a causal pathway but as another way to demonstrate the link between unfit parents and unfit progeny, a basic plank of eugenic thinking. Opponents within the medical temperance movement, however, found Pearson and Elderton's conclusions deeply disturbing and sought to rebut them. So vigorous was this rebuttal that, within the year, Elderton and Pearson saw fit to publish *A Second Study . . . Being a Reply to Certain Medical Critics of the First Memoir and an Examination of the Rebutting Evidence Cited by Them.*[77] The longer title reveals the evolution of the debate, which by then hinged not only on the substance of Elderton and Pearson's claims but on their way of knowing. Physicians, or "certain medical critics," sought to discredit the Galton lab's statistical findings and thereby to preserve medicine's prerogative to postulate about alcohol and heredity.

Elderton and Pearson began their first monograph, a statistical analysis of data on some three thousand schoolchildren in Edinburgh and Manchester, by claiming,

> An attempt is made in this paper to measure the effect of alcoholism in the parents on the health, physique and intelligence of their offspring. The question of intemperance is one of the chief problems of our national life, and as such, it is beset with difficulties. It is surrounded with prejudices and has been too often treated with rhetoric, so that it is extremely difficult to free the mind from preconceived opinions and approach the subject with a purely judicial and calm statistical spirit.[78]

By associating their statistical methods with a "judicial and calm spirit," Elderton and Pearson presented themselves and their methods as free of ideological fervor. (Whether their pretense to be free of ideology is borne out is another matter.)

In fact, Elderton and Pearson introduced the notion of effective statistical control into the investigation of alcohol and heredity. "It is perfectly idle to

quote statistics of the number of the insane or of the imbeciles who have drinking parents, if at the same time the numbers of normal individuals who have
drinking parents are not cited."[79] They argued in their first study that "alcoholism in the parent may, like insanity, be the somatic mark of a defective germ
plasm in the stock. The child is defective not because the parent is alcoholic,
but because it is the product like the parent of a defective germ plasm."[80] They
argued essentially that some factor, some hereditary weakness or taint, both
predisposed the parents to alcoholism and gave rise to deficiencies in the offspring.

> No *marked* relationship has been found between the intelligence, physique or dis-
> ease of the offspring and parental alcoholism in any of the categories investigated.
> On the whole the balance turns as often in favor of the alcoholic as of the non-
> alcoholic parentage. It is needless to say that we do not attribute this to the alco-
> hol but to certain physical and possibly mental characters which appear to be as-
> sociated with the tendency to alcohol. (32)

Although Elderton and Pearson were principally interested in eugenics,
their findings had obvious and unsettling implications for the temperance
movement. Elderton and Pearson voiced the heterodox opinion that a child's
development may be adversely affected by external environment rather than
by heredity alone. They speculated that "Alcoholism in the parent may have
no hereditary origin . . . Alcohol may be the source of evil to the children, not
because of physical changes wrought in the parents, but because of economic
and moral changes produced in the home environment" (2). Elderton and
Pearson directly challenged the ideology of the temperance movement by asking, "Would it not be of considerable importance to realise that we must look
to other factors besides drink as potent sources of our social difficulties?" (2).
Their conclusions were deeply disturbing to many temperance thinkers. The
temperance movement was heavily invested in the notion that the bodily weakness of alcoholism was a societal scourge as well. To locate the evils of intemperance outside the body, as Elderton and Pearson argued, was a radically dis-
junctive challenge to the primacy of alcohol as a social poison, as well as a
somatic and germ poison.

Indeed, the first Galton Eugenics Laboratory study threatened some of the
tenets dear to the advocates of temperance, particularly those in Great Britain,
who were quick to formulate critiques of the first study. But the monograph

garnered much attention in the United States as well. One of the most serious charges was that Elderton and Pearson had used moderate drinkers, rather than complete abstainers, as controls. Moreover, so the critics charged, they had failed to ascertain whether parents' drinking preceded the birth of the child.[81] This was a key dimension in the eyes of the temperance advocates, since germ damage could be attributed to drinking that occurred before the child's birth, as suggested by Forel's theory of blastophthoria.

However, one of the main objections to Elderton and Pearson's study seems to have been that they were not physicians and thus lacked the authority to rule on the matter of alcohol and heredity. As one correspondent wrote, "As a matter of handling statistics, Professor Karl Pearson seems to have completely answered his critics, yet no one with much practical medical knowledge can doubt that he is wrong in his conclusions."[82] Critics accused Elderton and Pearson of being "mere mathematicians" and of refusing to accept medical aid in their work. "On this 'if' and on these two 'somes' Professor Pearson and Miss Elderton have based a mass of figures and mathematical curves which they presented to the world as a model of scientific caution and accuracy, the alleged absence of which in other inquiries they assert in admonition," charged a letter from two physicians in the venerable *British Medical Journal*.[83] In part, the physicians' rejection of Elderton and Pearson's findings was an ideological one: they disagreed with the results, so they took issue with the methods. However, it also may have reflected physicians' unfamiliarity with this kind of abstract, quantitative evidence. Temperance advocates found "the purely judicial and calm statistical spirit" of the Galton pair unconvincing, when compared with the immediacy of case studies and personal experience — or ideological fervor — in shaping knowledge (though certainly Pearson and Elderton were no less ideological). In other words, new ways of knowing precipitated radical shifts not just in the form of knowledge but in its content as well.

Elderton and Pearson, for their part, countered that "The intellect of man can be affected by toxicants more subtle than alcohol, and the most dangerous of these is passion for collecting, without weighing, any statements which will support a prejudgment."[84] Thus, while staking a claim for the appropriateness of statistics to settle social questions, they in turn accused physicians of misusing statistics and misunderstanding statistical techniques. The ferocity of the debate over the Galton lab monographs was shaped not just by ideological differences in the ways social observers thought about heredity. It was driven as well by an argument over whether statistical methodology or clinical acumen was more "accurate" and thus who had authority to answer the question of alcohol and offspring: social science or medical science.

In many ways, Elderton and Pearson's reports and the heated response to them re-ignited the smoldering debate on the role of alcohol in racial degeneration or improvement. This debate largely pitted temperance thinkers, who saw alcohol — because of its damaging effect on individual lives — as an unmitigated deterrent to human progress, against a small group of eugenicists, who viewed alcohol ultimately as a means to improve racial stock.

The controversy over the Galton memoirs also reenacted the battle over the roles of nature and nurture. The debate about the role of environment in relation to heredity reverberated throughout nineteenth-century thinking. The physicians who objected so strenuously to Elderton and Pearson's findings sought to locate social ills in "nature," in the hereditary transmission of weak constitution from one generation to the next. In contrast, Elderton and Pearson proposed that "nurture" — the disordered home environment of the children of alcoholic parents, the "poverty, disorder, misery in which they are commonly brought up"[85] — explained the poor outcomes of these children. The debate revealed a deeper schism in social thinking. Because of the social and ideological usefulness of the notion that acquired characteristics were heritable, hereditarians clung to this idea. In fact, hereditarianism was both an important explanation of social disorder and a crucial lever for social change. Hereditarians often invoked the biblical metaphor of the sins of the parents being visited on the children; thus, the doctrine served both as a moral injunction and as a means to satisfy the quest to explain the observed world.

Alcohol as Race Poison

As the eugenics movement intensified, the notion of alcohol as a poison was again transformed metaphorically and literally. The metaphor of alcohol as a germ poison held explanatory power beyond notions of heredity. In the last decades of the nineteenth century, temperance advocates and eugenicists alike began to speak of alcohol as a "racial poison," a "substance which, whether or not injuring the individual who takes it, is liable to injure the race of which he is the trustee."[86] This metaphor drew not only on the concept of alcohol as a germ poison but on the idea of "race suicide," a preoccupation of the eugenics movement. In the early twentieth century, *race* was used to denote a tribe, nation, or people regarded as being of common stock — for example, the Northern European race, the German race, or the British race. Many eugenicists believed that certain stocks were committing race suicide by failing to reproduce rapidly or even at replacement level. Demography, the new science of fertility measurement, made such fears manifest; indeed, the notion of race suicide implicates women's fertility. The racial poison of alcohol became another element in the social construction of race suicide. As eugenics grew, concern about alcohol moved beyond its effect on individuals and their families to its effect on an entire race, or stock of people. The threat of alcohol was magnified from germ poison to race poison. The idea of racial deterioration inflated Morel's theory of degeneration from the level of the individual and family to the level of the society and the race. As the alcoholic lineages demonstrated, degeneration sowed destruction not only within the generations of the family but also within society. Describing the families depicted in Figure 1, a neurologist argued, "One such family is capable of throwing into the community dozens of useless or dangerous individuals, who, if capable of multiplying, will produce their like. If by depopulation is meant loss of individuals not only in a quantitative but also in a qualitative sense, alcoholism is undoubtedly one of its causes."[87] As another physician noted, "If the mother as well as the father is given to drink, the progeny will deteriorate in every way, and the future of the race is imperiled."[88]

Yet this contention was not shared by all. In fact, there were two distinct opinions about alcohol's "racial" effect. On the one side, most physicians, temperance advocates, and eugenicists saw alcohol as a racial poison, weakening society over time through the process of degeneration. Of course, this was the belief of temperance advocates who sought to make alcohol's detrimental effects on race and nation another plank in their campaign against drink. Prohibitionists claimed to be moved less by the individual harm inflicted by alcohol than by the social damage, particularly as eugenics gained social influence. "It is not the individual poisoning which constitutes the chief menace of alcohol. It is the chronic and racial poisoning which strikes at the root of future generations and lowers the level of citizenship," wrote Clarence True Wilson and Deets Pickett in The Case for Prohibition.[89] The concept of racial poison furnished temperance advocates with a new weapon in their battle against drink: even if the drinker did not harm himself through drinking, he must nonetheless stop drinking, as a "trustee of his race." In contrast, eugenicists who saw alcohol as a racial poison sought to eliminate not the drink but the drinker.

prohibitionists → eliminate the drink

eugenicists → eliminate the drinker

} both see alcohol as racial poison

Alcoholism rendered individuals unfit for parenthood; thus, it was sufficient reason to practice what the British physician C. W. Saleeby called "negative eugenics" — "the discouragement of parenthood on the part of the unworthy."[90]

On the other side, a contingent of social Darwinists, most prominently the British eugenicist G. Archdall Reid, adhered to a doctrine of selective improvement. They viewed alcohol as a "fool-killer," a positive evolutionary force that eliminated the weakest and least fit members of society. These eugenicists believed that "alcohol, like disease, is the cause of an evolution protective against itself."[91] To explain this effect, they hypothesized that individuals were vulnerable to alcohol but that races could become "immune" to its poisonous effect. Reid's contention that "drunkenness in the ancestry is the cause of temperance among the descendants"[92] was radically different from prevailing notions that inebriety caused hereditary taint, weakness, and intemperance in offspring. In fact, Reid's beliefs reversed the process of degeneration described by Morel. Reid's 1902 book *Alcoholism: A Study in Heredity*[93] epitomizes the anti-prohibition, pro-eugenics argument taken by this side in the debate. This book melded two of the major intellectual theories of the day: evolution and the germ theory. Espousing a Darwinian model of evolution and expressly dismissing Lamarckism and any possibility of the inheritance of acquired characteristics, Reid adapted germ theory and the new theory of immunity to the process of racial evolution in a social Darwinist mode. Moreover, Reid and his followers argued that alcohol may have different effects at the individual and the social level. As stated by Leonard Darwin, "Far from injuring the *race* (as distinguished from the *individual*), wholesale drunkenness has a beneficial rather than a harmful racial effect, by causing a wholesale elimination of degenerates and those lacking moral grit."[94]

Reid saw disease as an engine of human evolution. His theory rested on three "facts": that men have differing susceptibilities to disease, that offspring tend to inherit their parents' powers of resistance, and that disease is highly selective in its action. How Reid arrived at these facts illustrates the protean nature of knowledge, as we will see. An important corollary to his theory was the idea that "every race is resistant to every deadly disease strictly in proportion to its past experience of it."[95] Reid noted that populations adapted to local disease conditions: "the Negro flourishes in West Africa, where the Englishman perishes. The Negro dies almost as surely in London" (5). He argued, for example, that a race that has experienced tuberculosis over time is more, not less, resistant to it. "Deadly disease is therefore a selective agency. It weeds out the less resistant to it, leaving the race to the more resistant" (34). Over time, an entire race could acquire immunity to a disease. And just as races might acquire immunity to disease, so might they acquire immunity to alcohol, Reid argued.

Reid found strong support for his theory in the differing responses of human groups to alcohol, arguing that historical evidence demonstrated that the longer a people had been in contact with alcohol, the more "resistant" they became to it.

> Of all peoples, to whom alcohol is accessible, the most temperate are those who have been most afflicted by it, the least temperate are those who had little or no experience of it. *Every race, in fact, is temperate strictly in proportion to its past sufferings through alcohol.* Thus, the so-called savage races who have had the least exposure to alcohol are proverbially intemperate. They are furious drinkers, and are furious in their drink. Their intemperance frequently takes a more violent and homicidal form than is common amongst modern Europeans, and to that extent hastens their elimination. (97–98, emphasis in original)

In fact, Reid argued, "advanced civilization is equally impossible to all races which have not undergone as considerable an evolution against alcohol."

> The evidence on the subject is very clear. The Jews, Greeks, Italians, South Frenchmen, Spaniards, and Portuguese, who have longest had an abundant and continuous supply of alcohol, are the most temperate people on earth. North Europeans, and others who have had a less plentiful supply, are less temperate. Whereas some savages and others who have had little or no experience of the poison are the most intemperate.[96]

The idea of temperance or prohibition deeply offended Reid's eugenist sensibilities. Believing in alcohol as "a most potent cause of evolution," he argued that the only way to eliminate drunkenness from society was not to ban alcohol but to prohibit drunkards from reproducing. "It is in our power by copying Nature, by eliminating not drink but the excessive drinker, by substituting Artificial for Natural Selection, to obviate much of the misery incident with the latter, and thus speedily to evolve a sober race."[97] Ultimately, Reid proposed eugenics as the proper solution to the social ills attendant on drinking.

"Which is worse: that miserable drunkards shall bear wretched children to a fate of starvation and neglect and early death, or of subsequent drunkenness and crime, or that, by our deliberate act, the procreation of children shall be forbidden them?"[98]

These theories had particular sway in the American eugenics movement of the 1920s, especially in how American eugenicists viewed the parallel social movement of temperance. Reid's arguments persuaded some eugenicists to oppose Prohibition altogether and vehemently, contending that "alcohol has been an agent — possibly a very considerable agent — in improving man both in his physical constitution and in his intelligence and moral character."[99] One doctor questioned whether "in an effort to save the weakling from its abuse we are not preserving a degenerate strain that were better destroyed before it multiplies too far."[100] Indeed, Morel had argued that "It is a law for the preservation of the race, which strikes alcoholics with early impotence, and their descendants are not only intellectually feeble, but this degradation is joined with congenital impotence."[101] Another American eugenicist sought to reconstrue Prohibition as an issue not of temperance but of eugenics: "Prohibition is above all a Eugenical question . . . and a great social, political and biological mistake."[102] Far from seeing alcohol as a race poison, these eugenicists argued that "the elimination of alcohol is injurious to the race."[103] Rather than weakening society, alcohol strengthened it by attacking the weak stock. "The enlightened nations are the inebriate nations."[104] Such views privileged modern European and American societies; in the United States these beliefs linked the eugenics movement to the anti-immigration nativist sentiment that also prevailed in the early twentieth century.

Prohibitionists, in contrast, vigorously objected to the laissez-faire attitude of those, such as Reid, who put their faith in "nature's scheme of eliminating the drunkard."[105] They saw the eugenicists as embracing a brand of social fatalism, "the old theological doctrine of eternal damnation."[106] Temperance advocates disputed the notion that alcohol ultimately betters the human lot. "It is dangerous to argue that alcohol by killing off the weak ova or embryos or foetuses is benefitting the race; if every killing off of the weak is accompanied by a weakening of the strong the race will not gain but lose. Such an argument is really sounding the note of retreat in the midst of the advance of hygiene; the 'better dead' cry is an acknowledgment of defeat — it is the proclamation of pessimism."[107] Ballantyne refers in this passage to the pessimism of the social Darwinists, who saw no scope for planned social reform, preferring instead to let alcohol run its evolutionary course through human populations, weeding out the weak.

Indeed, Reid's theory of alcohol as a positive and potent force in human evolution proved to be a lightning rod for controversy. Temperance advocates were offended by his anti-temperance stance and the implications of his theory. Professor Simms Woodhead of Cambridge University, for example, objected that "I cannot hold for a moment that we must have a drunken nation before we can have a sober people."[108]

Prompted by theories of racial degeneration, both eugenicists and temperance advocates looked at alcohol consumption at the population level to advance their respective theories. "In view of the millions throughout the world who do not hesitate to drink alcoholic beverages, it is quite proper to ask whether the use of alcohol during past generations has actually injured the physical or mental quality of present generations. Or has its use actually benefitted or improved the quality of the present generations?" asked an editorial in the *Journal of the American Medical Association* in 1924.[109]

While prohibitionists attributed the demise of great civilizations such as ancient Babylon, Greece, and Rome to high levels of inebriety, others pointed to the superiority of the "drinking [i.e., white] races." In a congressional debate on prohibition in 1926, for example, one congressman pointed to the "handful of beer and ale drinking Englishmen holding in subjection 300 million Hindu teetotalers."[110] The prominent Cornell University scientist Charles Stockard claimed that "all of the dominant races have a definitely alcoholic history."[111] Dr. Henry H. Goddard wrote, "Indeed one must admit the argument that if alcohol did cause feeble-mindedness, the number of the feeble-minded would be enormously greater than it now is."[112] Similarly, Dr. Hobart Hare of Jefferson Medical College in Philadelphia maintained that "If it were true that alcohol always produces a habit to the extent of doing evil to all men, the population of the earth would have disappeared long ago."[113] These eugenicists held instead that both alcoholism and feeblemindedness were evidence of some form of taint. Other writers argued that alcohol acted as a racial poison only among the weak and that the ability to use alcohol moderately was a sign of "good stock." "It must never be forgotten in this connection that drunkenness, by shortening life, is continually purging the race of undesirables," noted a British doctor in 1917.[114] In other words, alcohol itself had beneficial eugenic properties. "Hard drinking has prevailed in the British Isles for the last 2,000 years, and yet it is safe to say that no army in the history of the world has shown greater virility than the army now doing battle for us in France and Flanders."[115]

Indeed, as these remarks suggest, the interest in alcohol's role in racial evolution and in determining the relative strengths of various racial groups gained greater urgency in the early twentieth century, as immigration and colonization increased and as World War I engulfed Europe. The 1915–16 volume of the *British Journal of Inebriety* contained seven articles on alcohol and the war effort. The eugenic view of alcohol lent more support to what the "enlightened nations" presumed the "natural" world order ought to be.

As the debate over racial degeneration versus improvement of the racial stock progressed, research on animals came to play an increasingly prominent role. Since it seemed impossible to settle the question of alcohol and heredity in human populations, researchers once again sought to introduce new kinds of evidence to the debate. The seemingly precise germ poison hypothesis about the effect of alcohol on offspring lent itself readily to investigation in the laboratory. Animal experimentation on the effects of germ poisons flourished in the first decades of the twentieth century: a 1931 review of animal studies on alcohol as a germ poison listed almost fifty published research reports.[116] The results of these experiments were widely discussed in both the temperance and the eugenics movements: they figured prominently in the discussion of alcohol as a race poison.

Yet laboratory science often operated in the service of social ideology, as illustrated in the animal experiments of Charles Stockard, an anatomist at Cornell with a lifelong interest in embryology and morphology. In 1909, Stockard began an extensive investigation of the effect of various toxins on embryonic development; beginning with minnow eggs, then progressing to hens' eggs, and finally guinea pigs, he investigated the effect of alcohol, ether, Chloretone (chlorobutanol), and other anesthetics on morphology. After his initial experiments, he focused his investigation on alcohol, not because he was particularly interested in alcohol per se but because it provided an ideal experimental material: he could easily administer it to his guinea pigs by vaporizing it and he could readily control the dose. In his early reports of these experiments, Stockard somewhat disingenuously denied any political or moral agenda. In a 1922 article "Alcohol as a Selective Agent in the Improvement of the Racial Stock," Stockard claimed that when he began his experiments in 1909 he was interested only in the "the modification of embryonic development" and that he had "no interest whatever in a sociological problem."[117] He asserted that "the fact that such deformities are experimentally produced in embryos by the application of alcohol does not in any sense demonstrate that human deformities have been induced by a similar cause . . . These results should be considered in an impartial manner and should not be distorted for one purpose or another in a discussion of the use of alcohol by man."[118] Yet this retrospective claim belied the nature of Stockard's own writing over the preceding decade. In 1916, with Europe disintegrating in war, Stockard had made grand claims about the role of alcohol in human evolution, based on his guinea pig experiments.

Stockard's initial findings on the descendants of alcoholized guinea pigs suggested that alcohol did have a detrimental effect on offspring.[119] He found initially that guinea pigs exposed to alcohol produced fewer and weaker offspring. However, as Stockard continued to follow the lineage of his alcoholized subjects, he found that in the fourth generation, the descendants of the alcoholized line were actually superior to those of the nonalcoholized guinea pigs. Thus, his conclusions changed as his experiments continued. He claimed that alcohol acted as a selective agent, hypothesizing that alcohol acted to kill off weak stock and thus produce superior guinea pigs in the fourth generation of the alcoholized lineage.

Stockard's findings were echoed by those of Raymond Pearl at the Johns Hopkins University. According to Pearl, his results with chicks demonstrated that "the racial effect of alcohol is preponderantly beneficial" and that this beneficial effect "appears to be the result of the fact that alcohol acts as a definite, but not too drastic selective agent, both upon germ cells and developing embryos, eliminating the weak and leaving the strong."[120]

Such findings fueled the debate over the evolutionary role of alcohol in human beings. Stockard did not hesitate to extrapolate from his animal findings to human society, concluding one of his studies on guinea pigs with the assertion that "Those nations of men that have used the strongest alcoholic beverages through many generations have now, from a standpoint of performance and modern accomplishments, outstripped the other nations with less alcoholism in their history."[121] Although Stockard was concerned with eugenics, his comments must also be heard against the backdrop of the ongoing national debate over Prohibition. Would Prohibition leave the nation weaker rather than stronger? Stockard's alcoholized guinea pigs seemed to suggest so.

In fact, Stockard came to hew to a strictly eugenist line of reasoning, believing in the eugenic power of alcohol at the population level.

Should one desire to apply these experimental results to the human alcohol problem, it might be claimed that some such elimination of unfit individuals had benefitted the races of Europe, since all of the dominant races have a definitively alcoholic history, and the excessive use of alcohol was decidedly more general three
or four generations ago than it was today . . . The great problem in preventative
medicine may some day be the prevention of physical and mental, or structural
defectiveness. And the actual prevention of such conditions is the elimination of
the germ plasm from which they arise and the preservation of the biologically good
stock.[122]

Mothers of the Land: Social Order and Female Responsibility

By the early twentieth century, in the eyes of both prohibitionists and eugenicists, the stakes were higher than ever. The debate no longer focused on the
effect of alcohol on individual offspring and families but rather on nations and
races. The future of civilization was at stake. Even as eugenicists focused their
debates on the role of alcohol in shaping the fate of races, women as individuals came under increasing scrutiny as well. The question of where to locate
responsibility for deficient children and for the fitness of future generations was
an important undercurrent in late-nineteenth- and early-twentieth-century
thought, just as it is today. Who was culpable for an imperfect child: the mother
or the father? The parents or the society? Before the mid nineteenth century,
notions of alcohol's effect on offspring saw both father and mother as determining the hereditary endowment of the child, with the paternal influence receiving more attention. Recall the case studies of T. D. Crothers: although the
drunkenness of both parents played a role in the habitual inebriety of his patients, the father in each case was especially culpable. The greater emphasis
on the paternal role in the hereditary effects of alcohol was a reflection of the
primacy accorded to the male role in reproduction and society, as well as the
greater social latitude permitted men, their consequently greater tendency to
drink and be drunkards, and the social assumption that most drunkards were
men. In contrast, notions of femininity constrained women's drinking to a
much greater degree. Nevertheless, by the last quarter of the nineteenth century the maternal influence took on increasing prominence in explanations of
heredity, as the cult of domesticity waxed in Victorian England and the United
States.

Female responsibility for children — and, by implication, for society itself—had been escalating over the course of the nineteenth century. Certainly the mother had always been considered an important influence on the off-spring; yet what is noteworthy about the last years of the nineteenth century is the shift towards thinking of the woman as primarily if not exclusively "responsible" for reproductive outcome. Moreover, female responsibility extended beyond children and family to the nation or race. Women were "mothers of the land. Upon the health of their bodies, and upon the integrity of their moral and spiritual nature, depends not only the welfare of their children and their homes, but the still larger and more serious extension of the family which we know as the nation."[123]

As society focused more attention on maternal responsibility, doctors began to locate the harm of alcohol in utero rather than in the germ plasm, as a germ poison. By the 1880s, doctors had begun to recognize an effect of alcohol in utero, as distinct from a hereditary effect: "The influence of alcohol on the unborn infant is distinct from the question of racial poisoning or blastophthoria . . . alcohol may reach the developing child via the mother."[124] As early as 1883, the *Maryland Medical Journal* carried an article titled "The Effect of Alcohol upon the Fetus through the Blood of the Mother," and in 1888, the Edinburgh obstetrician J. Matthews Duncan had systematically noted the higher incidence of abortion, miscarriage, and premature labor in "soakers."[125] Physicians believed that the prenatal effect of alcohol took two forms: a higher incidence of spontaneous abortions and stillbirths and of enfeebled and mentally defective children in pregnancies that came to term.

Physicians had also begun to distinguish this in utero effect from the hereditary one that had held sway for so long. "It is true that if a woman keeps in a besotted state during the entire period of gestation the tissues of the child *in utero* are necessarily injured; but the injury here is one acquired by the developing child: it is in no sense an inherited one," noted Leonard Darwin.[126] By the end of the century, physicians had begun to speculate—in language that powerfully prefigures the writing on fetal alcohol syndrome nearly a century later—about the "intra-uterine alcoholic environment"[127] and its effect on the fetus. Although physicians had previously associated spontaneous abortions, stillbirths, mental defects, and small birth size with alcohol, the notion that alcohol exerted an effect in utero was not seriously investigated until the late nineteenth century.

There is no way to know for certain whether physicians of a hundred years ago were seeing in their patients the cluster of symptoms that modern doctors might diagnose as fetal alcohol syndrome. However, by the last decade of the nineteenth century, doctors increasingly spoke of the effect of alcohol in utero rather than on the germ plasm. "Habitual intemperance on the part of the female when pregnant must tend to impair the development of the fetus *in utero* by impairing cell growth," contended an editorial in the *Quarterly Journal of Inebriety* in 1899.[128] That is, doctors had begun to think of alcohol as a teratogen acting during pregnancy rather than as a mutagen acting before conception. Consequently, they focused more closely on female culpability for damaged offspring. Certainly they were aware of the higher incidence of spontaneous abortion and stillbirth among alcoholic women, as recognized by medicine today. Ballantyne noted that "premature labors and stillbirths [were] common in the history of the women who drink hard during their pregnancies."[129]

Speculation about the in utero effect of alcohol was supported by new kinds of evidence. William Sullivan's 1899 study of women in the Liverpool prison exemplifies the new medical thinking about the relationship between alcohol and offspring. Sullivan's study was one of the first attempts to measure carefully the potential prenatal effect of alcohol. The study, published not in one of the medical temperance or eugenics journals but in the *Journal of Mental Science*, received widespread attention. Unlike most nineteenth-century observers, Sullivan avoided the methodological flaw modern demographers would fault as "sampling on the dependent variable" and included adequate controls. Typically, nineteenth-century studies investigated selected populations such as the institutionalized or prison inmates, identifying "degenerates" or imbeciles and then inquiring as to their parentage. Such a method fails to control for parental history of alcoholism in noninstitutionalized or nondegenerate populations. In contrast, Sullivan investigated, as he put it, "not alcoholism in the ancestry of the degenerate, but degeneracy in the descendants of the alcoholic."[130] He did so by comparing six hundred children born to 120 alcoholic women inmates in the Liverpool prison with children born to 28 nonalcoholic female relatives. He found that fetal and infant mortality rates were about two and one-half times higher in the children of alcoholics.[131]

Sullivan noted also that the children of alcoholic women incarcerated during pregnancy and thus unable to drink had higher survival rates than their siblings born when the mother was free to drink during pregnancy. He hypothesized about "a direct toxic action on the embryo, owing to continued excesses during pregnancy."[132] His speculations were echoed by many other doctors. J. W. Ballantyne, the author of the definitive *Manual of Antenatal Pathology and Hygiene*, published in 1904, contended that during the first trimester of pregnancy, which he characterized as the embryonic stage, exposure to alcohol could cause abortion or disease in the infant. In a paper of 1915 he noted, "Habitual soaking has a most dangerous effect upon both the life and health of the infant during its stages of uterine development, premature labors and stillbirths being common in the history of the women who drink hard during their pregnancies."[133]

There are some suggestive similarities between descriptions of the children of drunkards and our contemporary clinical picture of FAS. Doctors in the early twentieth century frequently noted the small size, small head, unusual facies, "deformities," and "idiocy" of children exposed to alcohol prenatally. They had begun to note characteristic congenital deformities in children of mothers who drank. For example, a Scottish obstetrician described the children of "female soakers" thus: "In reality these individuals are ordinarily small, sometimes large and thin, almost always incompletely developed, they present in general the attributes of the state known as infantilism. The cranium and upper part of the face are symmetrical, and sometimes there is a hemiatrophy of the whole body. The head is small, and the expression of the face strange and sad."[134] Duncan's observation of the "strange and sad" face of the children of inebriates is reminiscent of contemporary doctors' descriptions of the "funny-looking" faces of children with FAS. Keister likewise noted in 1911 the "peculiar looking anomalies" of children "born from degenerate mothers."[135]

Other observers often noted the weakness and sickliness of children of inebriate women, pointing out their "enfeebled organization at birth."[136] Such children had "impaired nerve vitality . . . signs of mental deficiency and lack of normal control . . . warped or stunted intelligence accompanied by impulsive, uncontrolled actions."[137] Crothers likewise described the physical defects of "children with degenerate ancestors." "They have faults of growth and structure and seem hypersensitive to the surroundings, or the opposite, dull and stupid, not recognizing sounds or being affected by adverse conditions."[138]

Although we cannot draw a direct parallel, these symptoms are certainly suggestive of those included in the official diagnosis of FAS today, particularly the mental retardation and "poor impulse control, problems in social perception, deficits in higher level receptive and expressive language, poor capacity for abstraction or metacognition, specific deficits in mathematical skills, [and] problems in memory, attention or judgment."[139] These earlier physicians may very well have been describing what clinicians today would recognize as fetal alcohol syndrome.

One reason for increased medical speculation about the prenatal effect of alcohol as distinct from the earlier focus on alcohol as a germ poison was that pregnancy itself became the focus of renewed attention at the turn of the century. Early in the twentieth century, the proliferation of articles on the perils of alcohol in infancy reflected the increasing attention to infant mortality. In campaigns to reduce infant death, doctors and public health advocates warned against drinking by nursing women and the practice of giving infants and children alcohol in the form of various medicines to induce sleep. Physicians and child-savers also emphasized prenatal care, often suggesting that a "good pregnancy" alone was enough to save a child. The prenatal period seemed to loom larger in their minds than the environment in which the child was reared, as if the nine months of pregnancy mattered more than the poverty, malnutrition, disease, poor sanitation, and overcrowding that the child might face after birth. Doctors thus tried to extend their control over pregnancy and pregnant women. In a 1922 symposium on infant mortality, Dr. Louise McIlroy "had not the slightest hesitation in saying that if she were given the responsibility of looking after the children of the country, she would rather surrender the responsibility of looking after the children after they were born than give up her control of the conditions for the nine months previous to their birth . . . To her mind, foetal welfare and foetal mortality were far more important subjects than infant welfare and infant mortality."[140] That is an extraordinary statement of the risk often invested in pregnancy in the early twentieth century. Pregnancy, always a risk-laden state, acquired a new level of riskiness and responsibility around the turn of the century, even as maternal and infant mortality rates were falling.

Moreover, physicians were quite explicit about the dangers of drinking by pregnant women and its sequelae. "The use of alcohol by pregnant women very materially and detrimentally affects the fetus. French authors have shown in the case of pregnant women who use alcohol that it, as such, passes in considerable quantities into the fetus and that non-developments, monstrosities and malformations are brought about in the alcoholized fetus. It is a well-known fact that abortion and miscarriage are frequent in inebriates."[141] Alcohol exposure during pregnancy not only left offspring with a weakened constitution or a susceptibility to the drink crave but caused "peculiar looking anomalies . . . filling our anatomical museums with these half-human monstrosities."[142] Around the turn of the century, the attention paid to the congenital deformities associated with prenatal exposure increased markedly.

As the prenatal influence of alcohol on offspring became more prominent, the paternal influence was eclipsed — indeed, even denied. "The disease is much more dangerous in the mother than in the father," claimed an 1893 marriage manual.[143] Sullivan, unlike many of his predecessors, believed that "the influence of maternal drunkenness is so predominant a force that the paternal factor is almost negligible."[144] By negating the male role, Sullivan placed the weight of the blame on the woman. A similar pattern was evident in the animal research on alcohol as a germ poison. Stockard concluded that the effects of alcohol in his guinea pigs — and by extension in human beings as well — were greater in females than males, suggesting that females were both more vulnerable to the damage caused by alcohol and more potent in transmitting that damage to offspring. In his reports on his own and Stockard's experiments with alcohol, Casper Redfield concluded that "the organism develops a resistance to the action of alcohol and this resistance is transmitted as a secondary sexual character to males, but not to females."[145] Likewise, T. T. Morton noted the "greater potency of inheritance through the mother, which seems to be acknowledged by all observers of alcoholic degeneracy."[146] In both guinea pigs and human beings, the female of the species was held to be peculiarly and singularly liable for reproductive outcome.

However, doctors were less clear about the mechanism underlying the fetal effect, speculating that it involved maternal blood "tainted" by alcohol "poisoning the placenta."[147] Once again invoking alcohol as a poison both literal and metaphorical, doctors hypothesized a variety of ways that alcohol might affect the developing fetus. As the notion of maternal impressions receded, doctors increasingly recognized the importance of the physical connections between pregnant woman and fetus. It was no longer a fright or shock or craving — an emotional event — that marked the child in the womb, but rather a physical effect of the mother's body. "While there is no nervous connection between the mother and the unborn offspring, the child may be subjected to the unfavorable influence of a poison through the blood stream," explained one doctor.[148] The unborn child was thus vulnerable to its own mother and her behavior during as well as after pregnancy.

In fact, much of the literature on drinking during pregnancy in this period presented subtly violent, highly transgressive images. To speak of mothers "poisoning" their own children is to reject what had previously been assumed to be the natural order of maternity. "How could a mother soaked in alcohol [during pregnancy] be expected to bear a healthy child?" lamented one physician.[149] In these constructions of the maternal influence on the fetus, women are seen to abandon their traditional nurturing role. Images of soaking and saturation, of besottedness, blood, and poison, characterized medical understanding of the prenatal influence of alcohol at the turn of the century. "I think it is not an exaggeration to state that alcohol is a poison, and that the foetus of a chronic alcoholic mother is itself a chronic alcoholic, absorbing alcohol from the mother's blood and subsequently from her milk."[150] The child was not just made in the mother's image but was, like her, a chronic alcoholic, even in the womb. But the mechanism was now seen to be not a matter of heredity but a matter of poisoning — and furthermore, poisoning by the mother, who ought properly to be the child's most devoted caretaker.

Yet even as the mechanism of the effect was physicalized and new teratogenic effects were imagined, the influence remained in many ways a moral one. In other words, the woman's drunkenness affected her child not just through the poisoning of the placenta, the alcoholic soaking, and saturation in utero, "the stamp of inebriety on the delicate tissues"[151] of the developing brain, but also through the moral degradation that inebriate women embodied. As one physician explained, "There can be no doubt that the ultimate character and physique of a child is greatly affected by the mental and physical condition of the mother."[152] As the inebriate mother soaked her unborn child in her own alcohol-tainted blood, she also soaked it in a morally degenerate, unnatural environment. "I think all men will acknowledge that woman is the divinely appointed guardian of the morals of the family and of society. We must acknowledge even more than this, which is that woman is the moulder of character as it is developed in the minds of the child and the youth."[153] As divinely appointed molders of character, women were expected to lead morally impeccable lives. They embodied a sacred duty to preserve society and uphold its most cherished ideals.

Attention to material inebriety during pregnancy coincided with broader and potentially disquieting social trends, particularly changes in gender roles and gender relations; changes in the way physicians thought about maternal inebriety accompanied significant shifts in women's social roles. Alcohol became the focus of prenatal campaigns just as women were breaking the bonds of Victorian domesticity.

Sometimes the medical literature on alcohol and reproduction explicitly addressed rapidly changing social mores. There were, for example, frequent references to the disturbing "increase of the drinking habit"[154] among women. "It is becoming quite a custom of late years among our so-called society women to keep a 'whiskey book' (a bottle shaped to appear like a small book), accessible when starting out on automobile spins, social calls to bridge and card parties, etc., in order to fortify them and refresh their conversational powers for these exasperating ordeals."[155]

Whether women's drinking was actually increasing or was simply receiving closer scrutiny than it had in the past is hard to say. Many observers claimed that women were drinking in greater numbers; moreover, these observers focused especially on the increase in drinking among women of the upper classes. "Intemperance is said to be spreading amongst women, not only of the lower classes, but also in the higher circles of society."[156] Increased drinking among women of the "wealthier classes" was particularly worrisome, because these women were presumed to hold the future of society. "We all have had experience of gin-drinking among the poorer class; it will be interesting to watch the results of the case of the wealthier classes where drinking of cocktails is so much on the increase. Will this affect the mental and bodily condition of the future generation?"[157]

However, far more radical changes in women's roles were occurring during this period. Two significant trends coincided with the greater emphasis on the maternal role in determining reproductive outcome: attempts to make higher education more accessible to women and efforts to increase women's control of the timing and spacing of births through contraception.[158]

Women's intemperance was viewed as an even greater threat to social order than men's drinking, which was widely acknowledged to be at the root of crime, poverty, and assorted vices. Women's inebriety caused less damage in the individual drinker than did male inebriety (the female drunk was believed to be less noxious than the male drunk), as well as less crime and poverty in the present, but it posed an undeniable threat to future social order. "On no class of women does the penalty for excessive use of alcohol fall with such disastrous effect as on those who are of childbearing age";[159] the disastrous effects of inebriety are manifested not in the women themselves but in the children they conceive and bear, and thus consequently in the society. Indeed, women bore social as well as individual responsibility; the shape of future generations lay in their wombs. Concurrent with this development of greater attention to the prenatal effect of alcohol was the shift towards a view of women as being solely responsible for progeny.

The increasing attention to what Sullivan called "the social gravity of female inebriety"[160] highlighted these transgressions from women's traditional roles as mothers and caregivers. In pursuing education, or even reproductive control, women seemed to be abandoning their "transcendent responsibilities"[161] to bear children for their family, their society, their race. Such women ignored "the instinct of motherhood, which puts self in the background, and gives all she can offer to her unborn child." They were "untrue to the duties of motherhood."[162]

With increasing frequency in the last years of the nineteenth century, physicians claimed that women were especially vulnerable to the ill effects of alcohol and thus especially deserving of moral opprobrium for their use of alcohol. In drinking alcohol, women endangered not only themselves but also their children and, by extension, the future of society as well. "In suppressing the female drunkard," Sullivan argued, "the community not only eliminates an element always individually useless and constantly liable to become individually noxious; it also prevents the procreation of children under the conditions most apt to render them a burden or a danger to society."[163] As women sought expanded opportunities, more education, and increased control over reproduction, physicians attempted to exert control over the roles permitted to women and to enforce their traditional roles as bearers of future generations.

> To sociologists it is evident that the question of maternal inebriety is one of the national importance, for many women fail to realize that alcohol in small quantity, if taken every day, may have very serious results. Any community which is sincerely interested in the general efficiency of its members will be sure to safeguard the health of its women, for these, on account of their maternal functions, may either become the resuscitating and repairing element of the race, or else may provide many of the elements of deterioration which are so greatly to be dreaded, and which are, to some extent, in evidence at the present time.[164]

Unconstrained, a woman was a threat to the social order. She must be safeguarded, her sphere carefully bounded, so that she might maintain her role as "resuscitating and repairing" the world and her own society.

Medicine and Social Order

Like all kinds of knowledge, medical knowledge is "inevitably differently formed in different social and historical settings."[165] Beginning with the Gin Epidemic, attention to alcohol's effect on offspring arose during periods of social ferment and amidst more general concern about the social effects of widespread alcohol use. Answers to the question of alcohol and offspring in the late nineteenth and early twentieth centuries were firmly situated in and expressive of beliefs about the social order and fears of disorder. Knowledge was shaped by debates about responsibility, cause and effect, lineages, statistical studies, natural experiments, and animal experiments, and it formulated a variety of hypotheses about the relationship of alcohol to heredity.

Ideas about alcohol and heredity began to coalesce and become systematized in the eighteenth and nineteenth centuries. Temperance advocates had long noted "the social blight of widespread alcoholic indulgence."[166] The evils of the hereditary effects of alcohol added another dimension to their concern. "If personal drinking of alcoholic liquors is injurious to health — and the facts show that this is true — it strikes at what is fundamental in society. It is therefore no longer a private matter, but a burning public question. Its consequences are passed on to others both by personal association and contact, and by transmission, thus entailing a burden upon society in the future."[167]

During the nineteenth century, society had come to believe that alcoholism in the parents could affect children in a number of ways: it could cause feeblemindedness and idiocy, and it could predispose offspring to drunkenness. And the consequences of alcohol could be transmitted in a number of ways: by the parents' drunken state at the time of the conception, through the inheritance of acquired characteristics, by damage to the germ cells, by alcohol acting as a racial poison, and eventually through fetal exposure to alcohol during pregnancy. "The problem of heredity, by which is meant the transmission of a parental and ancestral character to each new generation of organic beings, is one of transcendent interest in biology at the present time, not only because it seems to hold the key to a large part of evolution, but on account of its relations to many social, moral, and even political and religious questions."[168] Indeed, by the end of the nineteenth century, hereditarian explanations were applied to almost every visible social problem, not just intemperance.[169]

As explanations for the relationship between alcohol and offspring shifted, so too did concepts of risk and responsibility. From the early nineteenth century on, reproduction was viewed as a social act with significant social consequences. Thus, parents owed responsibility not just to their own progeny but to society as well.

> All intending parents should remember that they carry in their bodies the most precious of all earthly things, germ cells; and they should protect these from all evil influences — traumatic, toxic, toxinic, microbic — as they would their own lives, for by doing so they can give a great gift (none greater) to future generations. That in the act they are also benefitting themselves is surely an additional incentive to the less imaginative ones among them to whom posterity seems to matter little and race welfare to be an irrelevant consequence.[170]

Women increasingly bore the blame not just for individual defective children but for social deterioration. Figure 2 presents a vivid example of temperance propaganda exhorting women to uphold the social order by abstaining from alcohol. Yet the increasing focus on the prenatal effect of alcohol reflected not only advances in scientific knowledge but also tension over changing gender roles and aspirations. Just as women sought greater independence in both intellectual and reproductive spheres, physicians sought to enmesh them in an ideology of maternity that made them mothers both of their own children and of the society, the race, and the nation.

The configuration of knowledge about alcohol and offspring expressed the relationship between alcohol and the social order more generally. Alcoholism "leads to a degeneration not only of the individual, but also of the species; it is dangerous to society, as it produces a slow and progressive deterioration of the individual and an intellectual and physical sterility of the race with all its social consequences, viz., lowering of the intellectual status and depopulation."[171] Beliefs about alcohol, heredity, and social order were intertwined.

Regardless of whether doctors at the turn of the century were observing in their patients what doctors today would diagnose as FAS, there are some significant similarities between the construction of medical knowledge then and now. First, in both eras doctors have been concerned with the way private, individual behaviors have public, social impacts. Thus, drinking during pregnancy is worthy of sustained attention since it affects not just the pregnant

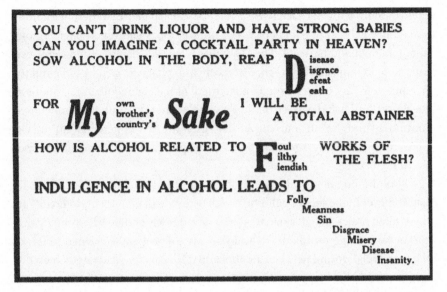

FIGURE 2. Anti–Saloon League billboard of 1909. From Andrew Sinclair, *Era of Excess: A Social History of the Prohibition Movement* (New York: Harper Colophon Books, 1962).

woman, nor even just the pregnant woman and her fetus, but the whole of society as well. Second, because of these social consequences of individual behaviors, doctors have always seen pregnancy as worthy of special intervention. They have seen it as their responsibility to protect future generations; they have acted both for the sake of the affected children and for the good of society. Third, doctors have used pregnancy as an opportunity to control women's behaviors and more generally to enforce particular gender roles.

Yet the knowledge so painstakingly and exhaustingly constructed during this period faded from medicine's collective consciousness in the mid twentieth century. The debate over prohibition was in many ways a culmination of nineteenth- and early-twentieth-century ideas about alcohol and heredity. It tested the concept of heredity, which had been unchallenged in the earlier period. By the beginning of the twentieth century, empirical research, however crude by today's standards, was increasingly brought to bear on the question of alcohol and offspring. The ideologies of the temperance and eugenics movements, along with the "scientific evidence" marshaled by different factions within the debates over eugenics and prohibition, determined later notions about alcohol and heredity. The years that were the "period of most enthusiastic hereditarianism" (1885–1920)[172] were also characterized by fervent temperance activism, culminating in enactment of the eighteenth amendment to the Constitution and passage of the Volstead Act in 1919. Both the temperance movement and the eugenics movement, itself the heir of hereditarianism, expressed the peculiarly American belief in the power of individual will to effect social change.[173]

Prohibition, along with the discrediting of eugenics after World War II, had the eventual effect of calming the debates about alcohol and heredity in the United States. What is remarkable is how quickly and how thoroughly this earlier knowledge vanished. Both physicians and laypeople seemed to forget the vexed relationship between alcohol and offspring; medical research on the question virtually disappeared, as well as public discussion of it.[174] As early as 1934, one year after Prohibition ended, in a book titled *Heredity and Disease*, Otto Mohr claimed that there was no sound evidence for the belief that alcohol and other chemicals cause heritable changes.[175] Eight years later, in 1942, doctors asserted that "no acceptable evidence has ever been offered to show that acute alcoholic intoxication has any effect whatsoever on the human germ, or has any influence in altering heredity, or is the cause of any abnormality in the child."[176] Thus, medical thinking had shifted rapidly and radically.

Because of this, the ways in which nineteenth- and early-twentieth-century physicians and laypeople attempted to answer the "question of alcohol and offspring" do not really serve as logical or even linear antecedents to our own ideas about an "alcoholism gene" or alcohol's effect on children, in particular the concept of fetal alcohol syndrome. The route from the knowledge possessed in earlier centuries to our late-twentieth-century ideas is not a progressive evolution of medical knowledge but rather a process that moves ahead in fits and starts, dredging up past concepts, discarding others, and inventing new ones. Nineteenth- and early-twentieth-century concepts of alcohol and offspring were constructed using the tools provided by contemporary thinking on heredity, racial degeneration, social order, and social welfare, including tools appropriated from the eugenics and temperance movements. Physicians of the period were observing behavioral outcomes and making sense of their observations based on the theoretical frameworks and tools at their disposal. Their answers to the question of alcohol and offspring thus reified their existing prejudices and beliefs.

The unending metamorphosis of answers to the "question of alcohol and offspring" demonstrates that medical knowledge is neither continuous nor necessarily cumulative; it is molded by human feeling, sociocultural forces, and historical context as well as by observed "fact."

CHAPTER 3

DIAGNOSING MORAL DISORDER

An amount of alcohol in the blood that would kill the mother is not enough even to irritate the tissues of the child. Part of the reason for this may be that alcohol is eliminated rapidly from the adult body; only very small quantities ever reach the child . . . Many publications claim to show that alcoholics give birth to more idiots, more malformed children, and more retarded children than do sober individuals. This claim simply does not stand up to scientific examination.

—Ashley Montagu

Given the medical attention paid to alcohol and reproduction in the nineteenth and early twentieth centuries, this statement by the popular and influential anthropologist Ashley Montagu in 1965 might seem surprising or even disingenuous. Yet Montagu's claim exemplifies the medical and scientific understanding of the significance of alcohol in pregnancy at that time. Despite the evidence that had been accumulating in different forms for 150 years, doctors in the middle decades of the twentieth century viewed the relationship between alcohol and reproduction very differently from doctors before or since. They showed little concern for the effect of alcohol on reproduction. The body of knowledge — facts, theories, experimental evidence, and moral conjecture — that medicine had accumulated on the subject of alcohol and reproduction seemed to fade from the collective medical consciousness. Morel's theory of degeneration was abandoned; Stockard's guinea pigs and Pearl's chicks were forgotten; physicians ceased speaking of alcohol as a race poison. Even the emerging theories about a specifically prenatal effect of alcohol were discarded. For several decades before the discovery of fetal alcohol syndrome, doctors believed that alcohol use in pregnancy was harmless and perhaps even beneficial.

The transformation of medical ideas about alcohol and reproduction in the mid twentieth century is readily apparent in the medical literature, medical textbooks, and doctors' recollections of this period. What is striking is how quickly medicine appeared to "forget" the connections physicians had earlier hypothesized between drinking and degeneration at both the individual and social levels. These ideas were not only forgotten, they were denied. For a period of roughly three decades at midcentury, doctors saw no harmful connection between alcohol and reproduction; in fact, they postulated a beneficial relationship in certain instances, even to the point of using alcohol therapeutically during pregnancy.

This shift in medical thinking was partly precipitated by changes in what constituted evidence and in what was considered appropriate methodology. As epidemiology became more sophisticated, physicians regarded the kinds of studies done by Bezzola and Howe in the nineteenth century as methodologically flawed. Crothers's case histories of alcoholic families seemed more quaint than compelling. Such studies were not the stuff of modern science. Moreover, medical theories of heredity and the understanding of genetics were rapidly evolving; Victorian notions of heredity seemed outdated. The entire eugenics movement was considered an embarrassment best forgotten, particularly after the horrors of the Holocaust.

Yet a more important reason for the transformation of medical thinking about the relationship between alcohol and reproduction lies in the broader social shifts occurring throughout the 1930s, 1940s, and 1950s, particularly a revolution in how people viewed alcohol and its role in society. With the repeal of Prohibition in 1933, the idea of temperance lost much of its moral sway. Drinking returned to fashion as a social and sociable activity; per capita alcohol consumption increased steadily after 1933, more than doubling in the next decade, from 0.97 gallon of ethanol per capita in 1934 to 2.07 gallons per capita in 1944.[1] Moreover, throughout the 1920s and 1930s it became more socially acceptable for women to drink. Whereas drinking had been a gender-segregated activity, now, increasingly, men and women drank together and incorporated cocktails into social life.[2] In this atmosphere, temperance advocates were marginalized and even ridiculed.

The glorification of drinking as a social activity is vividly reflected in the popular culture of this era, particularly in films such as *The Thin Man* series, in which the martini is a recurrent and favored prop. The cocktail era reached its apex in the 1950s, a time when the Rat Pack routinely performed with a bar on stage, mixing drinks between songs. Thus, in midcentury, alcohol shed the stigma of the earlier period and drinking became a much less stigmatized activity for men and women alike. And once Prohibition was repealed, academics and advocates reconceptualized the "drink problem," turning attention away from alcohol and towards alcoholism — the excessive and addictive use of alcohol by a minority of drinkers. Alcohol transmogrified from "demon rum" into "a safe, easily controlled anesthetic which acts on successive centers of the brain, notably the ones which govern man's higher intellectual activities," in the words of a Life magazine article.[3] Scientific attention focused on the "minority of drinkers who drank to excess." The focus on alcoholism was also stimulated by the formation of the self-help group Alcoholics Anonymous, which first met in 1935 and issued its first handbook in 1939.[4]

The medical literature of the late 1930s is largely silent on the matter of alcohol and reproduction. Animal experimentation on the effects of various toxic substances, including alcohol, on the germ plasm persisted until 1935 or so, but then researchers seemed to lose interest in alcohol as a mutagen. In the 1940s and 1950s there was very little focused attention on drinking and reproduction, and when alcohol was mentioned in conjunction with reproduction, doctors asserted the lack of evidence for an effect and the safety of drinking. Considering the volume of medical literature devoted to alcohol and heredity in the recent past, the silence on this topic in midcentury is astounding.

Within a year of Prohibition's repeal, doctors were claiming that there was no evidence for adverse effects of alcohol on heredity.[5] Less than a decade later, in 1942, the editors of the *Journal of the American Medical Association*, in a reply to a query about drinking during the early stages of pregnancy, felt confident in asserting that "in human beings it is difficult to prove that alcohol has a deleterious effect on babies *in utero*, even when large amounts are taken."[6] These claims reflected a double paradigm shift in medicine. In the first shift, physicians were relinquishing the old notions of drinking as moral failure and embracing the idea of alcoholism as a disease. In the second, they were repudiating the supposed hereditary effects of alcohol altogether.

Both shifts were visible in the work of Howard W. Haggard and E. M. Jellinek at the influential Center of Alcohol Studies at Yale University. The center evolved in the late 1930s and early 1940s and became a leader in the movement to recognize alcohol as a public health problem rather than a moral or criminal one. In contrast to the idea that alcohol was a social evil, these Yale researchers argued that alcoholism was a social problem and, moreover, that alcoholism was a *disease*, worthy of research and treatment.[7] The Yale center vigorously lobbied the American Medical Association to accept alcoholism as a treatable disease, a policy the AMA finally adopted in the 1950s. Haggard and Jellinek's popular 1942 book *Alcohol Explored* was widely read in medicine and exemplifies this twin paradigm shift. They contended that alcohol was a "social condiment" harmless to the vast majority of the population and posing a risk only to the unfortunate few who suffered the disease of alcoholism. They likewise repudiated the "poisoning" effect of alcohol on the germ plasm: "The fact is that *no acceptable evidence* has ever been offered to show that acute alcoholic intoxication has any effect whatsoever on the human germ, or has any influence in altering heredity, or is the cause of any abnormality in the child" (emphasis added).[8] With the leading medical authorities on alcoholism as a medical issue making such definitive pronouncements, it is little wonder that doctors considered the case against alcohol in reproduction to be closed.

This double paradigm shift contributed to the obscurity of older medical knowledge about alcohol and pregnancy. Modern physicians and researchers eagerly distanced themselves from their moralistic Victorian forebears. Indeed, earlier medical knowledge that saw alcohol as a potent germ poison was reduced to the status of mere folklore or myth: "One of the most persistent myths in the folklore of alcohol is that drinking injures germ plasm and has an evil hereditary influence, but the evidence does not bear this out; indeed, it is somewhat to the contrary. Alcohol perhaps assists sperm selection, for in weak concentrations occurring in the body fluid it seems to affect the feebler and less active sperms while not affecting the vigorous ones except to slightly reduce their motility."[9]

In stark contrast to the concerted childbearing and prenatal hygiene campaigns of the first two decades of the twentieth century, obstetrics textbooks of the 1930s and 1940s rarely mention alcohol in any capacity. In the 1950s and 1960s, references to alcohol and pregnancy appear with slightly greater frequency, though in these instances the focus of medical concern is entirely different. In this period, doctors were concerned only with the caloric load of alcohol and its potential to displace nutritional caloric intake. This statement from the 1953 edition of *Clinical Obstetrics* is representative:

Alcohol, as such, is not injurious and need not be eliminated during pregnancy. This, like many other things, should be undertaken in definite moderation. An occasional cocktail, highball, beer or ale, need not be restricted and may be most beneficial. However, alcoholic beverages such as beer, ale and mixed drinks have a relatively high caloric value. Probably the most serious effect of drinking these beverages comes from the calories they, and the hors d'oeuvres which usually accompany them, carry. Thus, they may increase the caloric intake above the proper level or the patient will eliminate items of required nutrients in an effort to keep her weight down.[10]

This view of alcohol — as potentially beneficial and at most harmful only because of its caloric content — persisted well into the 1970s. Both the 1970 and 1974 editions of *Synopsis of Obstetrics* claimed that "Alcohol used in moderation for social purposes apparently exerts no harmful effect on the course of the pregnancy or on the fetus. Alcohol may stimulate the appetite and is itself a source of calories and thus should be avoided by patients trying desperately to limit their weight gain."[11]

Throughout this period, doctors regarded drinking during pregnancy as a benign activity with no serious long-term consequences. According to the 1963 edition of *Textbook of Obstetrics and Obstetric Nursing*, "Cocktails, wine and beer are not contraindicated when used in moderation. A small glass of sherry before meals may stimulate an appetite. A glass of port before bedtime may help to promote sleep. It must be remembered that their caloric content is very high."[12] Again we find the warning about calories, but also an explicit endorsement of alcohol as an aid to appetite and to sleep. A 1973 edition of *Modern Obstetrics for Student Midwives* was quite liberal, by today's standards, in its stance on alcohol use during pregnancy: "Alcoholic drink taken in moderation has helped society for a very long time, aiding relaxation and sleep on the one hand, and promoting conversation and whetting the appetite on the other. If a woman has been accustomed to taking some alcoholic drink then there is no reason why she should forgo the pleasure merely because she is pregnant."[13] Ironically, this textbook appeared in the same year that FAS was discovered. As late as 1976, *Obstetrics: Essentials of Clinical Practice* acknowledges that "Alcohol readily crosses the placenta, but alcohol ingestion in moderation has not been shown to have an ill effect on the fetus. A large alcoholic intake, however, exerts a deleterious effect primarily because such a patient does not eat properly and may suffer malnutrition. An infant born of a chronic alcoholic mother may experience withdrawal symptoms during the perinatal period."[14]

As these passages suggest, the presumed benefits of alcohol in pregnancy included its reputation as a relaxant. In my interviews with doctors, some respondents recalled recommending alcohol as a relaxant when doing amniocentesis, the new procedure developed in the 1960s to test for genetic abnormalities in the fetus. One male obstetrician recalled that "We used to tell all of our patients to go home and have a glass of wine, 'cause it will keep their uterus from contracting after their amniocentesis. So we actually used to prescribe it" [26B].[15] Even in the FAS era, some doctors continue to recommend the therapeutic use of alcohol in pregnancy. "On occasion I still do say to a patient, if you have a lot of contractions and it's just irritability, have a glass of wine or some beer and that will cut it down for you," reported one obstetrician. Several obstetricians mentioned recommending a glass of wine to calm preterm contractions; one mentioned that fellow physicians were likely to administer this therapy to themselves during pregnancy. A female obstetrician noted, "We have a lot of physicians in the practice and, you know, sometimes when women are busy on their feet all day, they're pregnant, they have a little bit of what we call preterm contractions, and they'll say, 'Oh, I had a glass of wine last night. It knocked them out.' Because, of course, alcohol was used as a tocolytic agent" [28B].

This doctor's recollection of alcohol as a tocolytic — an agent to arrest preterm labor — recalls the most interesting shift in medical knowledge about alcohol and pregnancy in this period. In the 1960s, physicians began to prescribe alcohol therapeutically in pregnancy, to arrest preterm labor. Then as now, doctors had scant understanding of why some women seemed to go into labor early. And they were often desperate to find a remedy, since prematurity has well-known adverse consequences for the infant. Several obstetricians recalled the use of intravenous alcohol as a tocolytic to prevent or arrest preterm labor. "We used to use alcohol to inhibit premature labor and I remember how safe we used to think it was when we used to toss it around as if it wasn't a problem" [26B, male].

The procedure, which consisted of intravenous administration of ethanol (pure alcohol), originated with and was championed primarily by Dr. Fritz Fuchs of Cornell.[16] Fuchs developed the intravenous ethanol procedure when his own wife went into labor in her seventh month of pregnancy. Anna-Riitta

Fuchs was a physiologist who had found in her laboratory research that alcohol inhibited labor in rabbits. The Fuchses apparently decided together to apply the technique to Anna-Riitta, whose contractions stopped.[17] Subsequently, the Fuchses, together and individually, published more than thirty articles on the treatment. The use of intravenous alcohol to prevent premature birth was a common practice in American hospitals from the late 1960s until the early 1980s. Despite its apparent success in Anna-Riitta Fuchs's case, intravenous ethanol was in fact a brutal procedure for all involved; the women became extremely intoxicated and behaved accordingly. Doctors recalled that women "got so they smelled like a fruitcake." They also reported that nurses detested the regimen: not only was women's behavior obnoxious, but they frequently had to be forcibly restrained, and since nausea and vomiting almost invariably accompanied the treatment, there was the possibility of severe, even fatal, aspiration pneumonitis.

> The problem with it was, number one, it was very difficult to administer in the sense that the patient very quickly got blitzed and became uncontrollable very often, so that you would give IV alcohol and at the same time you had to strap her down, so you had to restrain her because she would become unmanageable. It was horrible from a nursing point of view. They, the nurses, hated it because of that aspect. And women who were treated once never wanted to be treated with IV alcohol again. [15B, male]

> I remember when I was a resident, there would be two or three or four patients on the floor getting IV alcohol. It was kind of comical because you had these staid serious women come in in preterm labor and we'd start the alcohol and then you'd have these major personality changes . . . We used to use it all the time, again and repetitively . . . It was funny because the whole room would smell like a bar. [26B, male]

Obstetrics textbooks throughout the 1970s and even into the early 1980s discussed this therapeutic use of alcohol. Both the language of the textbooks and the recollections of obstetricians suggest that doctors often used to recommend a less drastic form of this treatment, in the form of advice to women that they take a glass of wine or liquor to relax and to help calm early contractions. This therapeutic use of alcohol in the years just prior to the discovery of

FAS suggests that doctors no longer turned a wary, disapproving eye on drinking during pregnancy.

Between the mid 1930s and 1973, then, medicine utterly disregarded the "question of alcohol and offspring" that had so preoccupied it in earlier eras. Indeed, medicine came to regard alcohol in pregnancy not only as harmless but as quite benign, therapeutic even. Before 1973, medical beliefs about alcohol in pregnancy thus centered around four "facts." First, alcohol was not harmful, either as a germ poison or in utero. Second, the only reason to be concerned about drinking during pregnancy was the possible displacement of nutritive calories by the "empty calories" of alcohol. Third, alcohol could play a beneficial role as a relaxant for women who were to undergo amniocentesis or who were otherwise stressed. And fourth, alcohol administered intravenously was a potent tocolytic for the arrest of preterm labor. Both physicians and laypeople seemed to forget the vexed relationship between alcohol and offspring that had preoccupied their forebears; medical research on the question, as well as public discussion, virtually disappeared.[18] The concerns and social anxieties expressed in the nineteenth- and early-twentieth-century understandings of alcohol and heredity were forgotten in the mid twentieth century.

The remainder of this chapter explores the sudden reemergence of medical concern about alcohol in pregnancy in 1973, the social construction of the new diagnosis of fetal alcohol syndrome within the medical realm, and the early evolution of FAS as both a clinical entity and a social problem in the minds of doctors.

Creation of a Diagnosis

Our task in this chapter is to understand how a new syndrome,[19] with little definitive proof of its existence or its etiology, was first accepted as medical knowledge with a broad impact on public policy. The term *fetal alcohol syndrome* was coined by a group of specialists in dysmorphology (the study of human malformations) who published the first three articles on what they deemed "this tragic disorder," but the nature of the diagnosis and its salient symptoms were determined collectively over time by a loose confraternity of medical practitioners and researchers. FAS thus serves as a case study of the diffusion of new knowledge in the medical community, illustrating how a new diagnosis enters and permeates the medical consciousness. I analyze the first ten years of the

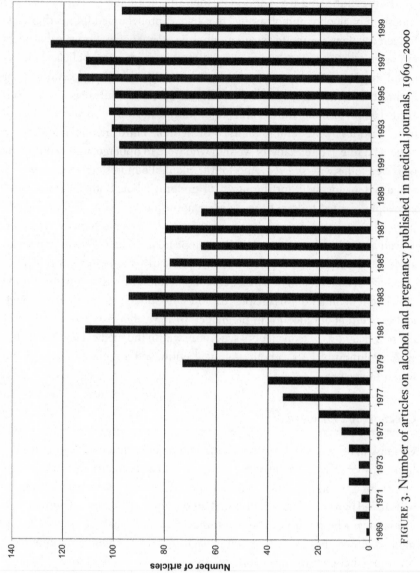

FIGURE 3. Number of articles on alcohol and pregnancy published in medical journals, 1969–2000

medical literature on FAS, concentrating on the construction of the syndrome in the United States — where it was discovered and where it has provoked the most concern[20] — and describe how new medical knowledge takes hold and becomes codified.

The recognition of a new disease or syndrome is sometimes the result of serendipity but more often the result of determined investigation and scientific entrepreneurship. In the case of fetal alcohol syndrome, the diagnosis emerged into medical consciousness at the intersection of social trends and events within and outside medicine. In addition, in the case of FAS, moral fervor powered the discovery as much as did medical curiosity. The doctors who discovered fetal alcohol syndrome were acting as moral entrepreneurs, in addition to being medical pioneers, on the frontier of diagnosis. A moral entrepreneur is someone who sees "some evil which profoundly disturbs him" and sets out to remedy the situation. Moral entrepreneurs are like religious crusaders, whose missions have "strong humanitarian overtones." They derive power not only from the legitimacy of their moral position but from their own "superior position in society."[21] That is, moral entrepreneurs tend to be elites working to impress their moral vision on the rest of society. In constructing a new diagnosis and, by implication, a social problem, doctors speak with unique authority: in our society, it is still primarily doctors who are granted the authority to identify a new disease.[22] The medical entrepreneurs who initially published their findings on FAS thus held a formal medical, as well as moral, mandate.

The medical literature serves as a "public arena"[23] in medicine. It is in this arena that new diseases, new theories, and new treatments must be vetted if they are to become part of the canon of medical knowledge. The medical literature also serves as a repository of medical knowledge and as a means of disseminating new knowledge to doctors around the country and around the world. Thus, the medical literature provides a window onto the process of knowledge formation in medicine. Discussion of the new syndrome was initially slow to take off (see Figure 3). There seems to be have been sustained but subdued interest in the subject for the first several years after the diagnosis was created. Then, in the late 1970s and early 1980s, the literature proliferated. The peak in 1981 is probably due to the announcement of the surgeon general's guidelines recommending abstinence from alcohol for all pregnant women and for those "considering pregnancy" and urging health professionals to inquire routinely about the alcohol consumption of their pregnant patients.[24]

Introduction of the FAS Diagnosis

Three articles published in the British medical journal *Lancet* over the twelve-month period from June 1973 to June 1974 are the foundation for the novel diagnosis of FAS. All three articles were written by members of the Dysmorphology Unit at the University of Washington School of Medicine in Seattle; these researchers constitute the core group of FAS entrepreneurs. Dr. David W. Smith in particular had an interest in identifying, naming, and cataloging birth defects and was the author of an atlas of birth defects, *Recognizable Patterns of Human Malformation: Genetic, Embryonic, and Clinical Aspects.*[25] The Dysmorphology Unit's identification of FAS was an extension of its work on chromosomal anomalies and part of the thrust in medicine at that time to categorize and name birth defects. Attention to birth defects surveillance intensified in the 1960s and early 1970s following the thalidomide disaster in Europe and the birth of babies affected by the 1964–65 epidemic of German measles in the United States. In fact, it was in 1973 that Virginia Apgar, of Apgar score fame, succeeded in establishing the national Birth Defects Monitoring Program.[26] The University of Washington has retained its status as the epicenter of FAS research and entrepreneurship for the last thirty years.

The first article on the FAS diagnosis, "Pattern of Malformation in Offspring in Chronic Alcoholic Mothers," was published in June 1973, with Kenneth Lyon Jones, David W. Smith, Christy N. Ulleland, and Ann Pytkowicz Streissguth sharing authorship.[27] (Three of these authors — Jones, Smith, and Streissguth — went on to devote major portions of their careers to research on FAS.) This article was presented as an early warning call to clinicians: "The purpose of this report is to alert physicians and other health professionals to a pattern of altered morphogenesis and function in eight unrelated children who have in common mothers who were chronic alcoholics during pregnancy."[28] It included detailed case reports of three Native American, three black, and two white children who shared a group of anomalies, including developmental delay, microcephaly, prenatal and postnatal growth deficiency, short palpebral fissures, epicanthal folds, small jaws and flattened midface, joint anomalies, altered palmar crease patterns, and cardiac abnormalities. The authors concluded their observations by claiming that "the data are sufficient to establish that maternal alcoholism can cause serious aberrant fetal development. Further studies are warranted relative to the more specific cause and prevention of this tragic disorder."[29] According to the authors' tabulations, the women in their study had been alcoholics for more than nine years, on average; more than half were known to experience delirium tremens; and at least a quarter had cirrhosis and nutritional anemia. In other words, their alcoholism posed severe threats to their own health.

The term *fetal alcohol syndrome* was first introduced in a second *Lancet* article, published five months later, by two members of the Seattle team.[30] This article described an additional three cases in Native American children, cataloged more of the children's shared anomalies, and named the disorder.

In June 1974 the Seattle team (composed this time of Jones, Smith, Streissguth, and Ntinos C. Myrianthopoulos) published a third *Lancet* article, "Outcome in Offspring of Chronic Alcoholic Women."[31] In this study, they used data from the Collaborative Perinatal Project of the National Institute of Neurological Disorders and Stroke. Using chart reviews, the authors identified twenty-three women with a history of chronic alcoholism and six cases of suspected FAS in children born to these women. Among the twenty-three women, perinatal mortality was 17 percent, 44 percent of surviving children had borderline to moderate mental deficiency, and 32 percent had abnormal physical features indicative of FAS. These twenty-three cases of maternal alcoholism and six cases of suspected FAS (none of the children was actually seen or definitively diagnosed) were found in a total sample of fifty-five thousand cases, yielding both a very low incidence of FAS and little evidence that heavy drinking during pregnancy was a widespread problem.

Intermittent reports before 1973 had noted a tendency for chronic alcoholic women to give birth to babies with a low birthweight, and a 1968 article in a French medical journal had reported on 127 children of alcoholic women,[32] but the three *Lancet* articles by the Seattle team were the first to delineate systematically and to label the association between chronic maternal alcoholism and a specific configuration of severe birth defects. Thus, these three articles — a total of eleven case reports of the syndrome and a noncontrolled, retrospective cohort study — formed the empirical foundation for the diagnosis of FAS. The process of building on that meager foundation occurred in several stages.

Confirmation and Corroboration of the Diagnosis

For the FAS diagnosis to have validity as a new syndrome, others had to replicate the Seattle Dysmorphology Unit's "findings." The group's observations

were quickly followed by a spate of articles, letters to the editor, and case reports describing newly recognized patients with FAS.[33] Like the initial *Lancet*
articles, these reports were based on case studies, typically small in number.
Often reports of just a single case were published, as in the December 1973 letter to the editor of the *Lancet* that began, "Sir, we think that we have observed
a case of the 'fetal alcohol syndrome' described by Jones et al."[34] Although
these articles primarily confirmed the original diagnosis, they occasionally
added new symptoms. For the first several years, authors reported an unofficial
tally of known cases, ticking up slowly from eleven[35] to twenty-six[36] to forty-
one,[37] and so on. This count continued for years after the diagnosis was first
introduced to the medical literature — the highest tally I could find in any article was 618 cases, in a 1979 article by David Smith.[38] In this stage of confirmation, documentation of individual cases still held a special status; the phenomenon was not yet documented or measured at the population level but
consisted rather of a gradual aggregation of individual cases. For example, one
article in *Early Human Development* proudly announced, "The world's literature has been scoured to obtain data on 492 examples of the fetal alcohol syndrome (FAS) in the Northern Hemisphere,"[39] a crude kind of meta-analysis.
Almost every author who published work on FAS in this period committed a
small act of scientific entrepreneurship by taking part in the process of collective determination that shaped the diagnosis of FAS.

Just as "a social problem exists primarily in terms of how it is defined and
conceived in a society,"[40] diagnosis exists in terms of how it is defined and conceived in the society of medicine. Although fetal alcohol syndrome was initially "discovered" by Smith and his colleagues, what FAS actually was — how
to recognize it in a patient, what symptoms it encompassed — was not determined solely by the Seattle team but rather was a process of "collective definition."[41] In other words, the diagnosis of fetal alcohol syndrome was shaped not
only by the underlying objective reality of alcohol-related teratogenesis but also
by how and what doctors chose to see in the syndrome. This selective and collective process of diagnosis definition is particularly evident for syndromes that,
like FAS, have no definitive biological markers. This is not to say that medical
classifications of disease do not correspond in any way with underlying biological and physiological "reality." Rather, diagnostic paradigms also inevitably
involve choices about what to include and what to exclude, and these choices
become more arbitrary, less definitive, when there is no biological marker and
the disease must be constructed by consensus in the medical profession. Com-

pared with Down's syndrome, which is readily identified not only by the distinctive "Down's face" but also by the presence of the chromosomal abnormality trisomy-21, fetal alcohol syndrome is a much more subjective, and therefore harder, diagnosis to make.

Diagnostic Dissent: Questioning FAS and Its Etiology

The establishment of fetal alcohol syndrome as a diagnosis was marked not only by consensus but by dissent as well. Early on, articles appeared that questioned the diagnosis. Some colleagues challenged Jones and Smith to define "severely chronically alcoholic."[42] Others questioned whether there was in fact a "similar pattern" of defects among the original eleven cases on which the diagnosis was based[43] or noted that the data were not controlled and were retrospective.[44] In letters to the editors of various journals, a serious debate ensued about the correlation of maternal alcoholism with other factors that put the fetus at risk, such as malnutrition, poverty, inadequate prenatal care, and smoking, and about whether alcohol itself is the causative agent.[45] Other authors noted that both the frequency and the severity of defects in the offspring seemed to be correlated with the severity of the woman's drinking problem, and one "hypothesized that embryonic disturbance is not as dependent on the amount of daily alcohol consumption as it is on the stage of maternal alcoholism."[46] Despite the resonance of this hypothesis with the original case reports of FAS, in which the mothers' acute alcoholism drastically threatened their own health, the medical literature continued to focus on any alcohol exposure in any pregnancy as risky.

A few articles attempted to address the puzzling fact that many heavy drinkers give birth to normal babies, postulating that an inability to process acetaldehyde, one of the by-products of ethanol breakdown, results in a particularly high blood concentration of acetaldehyde in the fetuses of some women.[47] Acetaldehyde is both cytotoxic and highly teratogenic. Yet despite the potential of this hypothesis to explain why only some heavy drinkers — and why even some moderate drinkers — have babies with FAS, it was quickly discarded.[48] Alcohol was assumed to be and accepted as the teratogenic agent at work.

Although this debate about the etiology of FAS appeared in the American medical literature, it was primarily and most vigorously conducted in the European literature. In the United States, the early debates on the etiology of what was called fetal alcohol syndrome followed a trajectory parallel to that demonstrated by Gusfield for drinking and driving: "In focusing almost exclusively on the factor of alcohol, [the debate] necessarily turns one element in a complex pattern of 'causes' into a single major factor."[49] In the case of FAS, the single-minded focus on alcohol as the sole cause of the observed outcome blinded doctors to the social context in which prenatal exposure to alcohol occurred and to any potential ameliorating or exacerbating factors. Every woman was equally at risk. But this assumption contradicted research findings suggesting that even among chronic alcoholics, not every woman would have a baby with fetal alcohol syndrome.[50] Moreover, some studies suggested that occurrence of the syndrome could be heavily influenced by other maternal characteristics. For example, in a study that examined women who had at least three drinks a day during pregnancy, the authors found a rate of FAS of 71 percent among children of low-income women but only 4.5 percent among women of higher socioeconomic status. The key difference between these two groups was their nutritional status during pregnancy.[51] More recent studies have hypothesized that alcohol acts as a teratogen only in combination with certain "permissive and provocative" cofactors, such as malnutrition, smoking, stress, or exposure to environmental toxins.[52] The focus on alcohol as the culprit in part reflects the conservatism that marked American culture of the 1980s; these were the years in which First Lady Nancy Reagan popularized the "Just Say No" slogan. But the focus on alcohol alone also stresses the individualization of this problem, by singling out drinking as the cause of FAS.

In fact, there was some debate about moderate drinking versus "alcoholism" as a risk factor for FAS. A letter to the editor of the *Lancet* questioned "why the syndrome remains rare while social drinking during pregnancy is so common," but doubts of this type were rarely seen in the literature.[53] Such deductive reasoning was largely ignored, and most of the skepticism was soon overwhelmed by claims that "the data supporting the presence of a fetal alcohol syndrome are enormous"[54] and "the existence of FAS can no longer be denied."[55]

Although dissent about the diagnosis of FAS was largely dismissed and had little impact at this point in its history, such dissent may be integral to the establishment of the disease in several ways. It is part of "the selecting and compressing process by which . . . facts emerge."[56] The very existence of dissent implies that the diagnosis was subject to professional debate and vetting; that such debate was ultimately fruitless may in fact impart a specious reality to the diagnosis. Even today, despite the emphasis on evidence-based medicine, dubious medical findings may be speedily and widely accepted within the medical community; it is difficult to extirpate such medical "knowledge" once it has taken hold.[57] In addition, the bias towards positive results in the scientific literature is exacerbated when there is an overt moral dimension to the research question at hand, as in the case of substance use in pregnancy. Studies of the scientific publication process have demonstrated a "bias against the null hypothesis" in the case of the effects of cocaine in pregnancy: studies showing no adverse effect were less likely to be reported than those showing a positive association, even though the latter were typically less sound scientifically.[58]

Widening the Diagnosis and the Scope of Expertise

Even before dissent about the existence and etiology of fetal alcohol syndrome began to recede, the diagnosis entered an expansive phase. During this period, the definition of what constituted fetal alcohol syndrome shifted. Doctors noted a variety of purported symptoms of prenatal drinking, some of which have become codified as part of the diagnosis, others not. However, what is most significant about this aspect of the discovery of FAS is not the particular symptoms, signs, and markers that came to be associated with the syndrome but rather the process through which doctors sorted out what *is* fetal alcohol syndrome. Over time, the symptoms associated with the diagnosis proliferated, so the social problem represented by the syndrome increased. In other words, although this expansion took place within the clinical realm, changes in the definition of the syndrome had wider ramifications as well. Shifts in the clinical perception of FAS precipitated an expansion of social concern about the threat represented by drinking during pregnancy. Such an expansion in the "domain" of a social problem often occurs after the existence of the problem has been established and accepted.[59] Once physicians believed that the phenomenon "fetal alcohol syndrome" existed, they began to "reconstruct" the diagnosis in a process that paralleled the evolution of social problems more broadly and that in turn contributed to the formation of the "new" social problem of drinking by pregnant women. Joel Best explains this process with respect to social problems:

> Initial claims-making must persuade people that a problem exists. Once these early claims gain acceptance and the problem becomes well established, with its own

Table 2. Sample of Articles Illustrating Diagnosis Expansion and Expertise Expansion

Author(s) and Date	Journal	Title	Claims
Møller, Brandt, and Tygstrup 1979	*Lancet* (letter to the editor)	Hepatic Dysfunction in Patient with Fetal Alcohol Syndrome	"Hepatic dysfunction has not previously been described in connection with fetal alcohol syndrome—perhaps it has been overlooked because of silent symptoms or perhaps it is a very rare complication of the syndrome."
Khan et al. 1979	*Lancet* (letter to the editor)	Hepatoblastoma in Child with Fetal Alcohol Syndrome	"Transplacental exposure to alcohol could have promoted liver cancer in this infant."
Steeg & Woolf 1979	*American Heart Journal*	Cardiovascular Malformations in the Fetal Alcohol Syndrome	"Pulmonary artery dysplasia is a unique deformity, but appears to be a common cardiovascular anomaly in this syndrome."
Qazi et al. 1980	*Teratology*	Dermatoglyphic Abnormalities in the Fetal Alcohol Syndrome	"The abnormalities of dermatoglyphics (the pattern of palmar creases and whorls) reported here constitute a valuable marker trait of the teratogenic effect of alcohol on fetal development and provide additional diagnostic signs for the fetal alcohol syndrome."
Cremin and Jaffer 1981	*Pediatric Radiology*	Radiological Aspects of the Fetal Alcohol Syndrome	"A survey of our cases has shown that skeletal maldevelopment is a part of the syndrome and can help in its early recognition."
Gonzalez 1981	*JAMA*	New Ophthalmic Findings in Fetal Alcohol Syndrome	"Such conditions . . . may be of help in diagnosing questionable cases of fetal alcohol syndrome."
Tredwell et al. 1982	*Spine*	Cervical Spine Anomalies in Fetal Alcohol Syndrome	"The occurrence of neck fusion in the fetal alcohol syndrome is common enough for it to be used in making the diagnosis."
Adickes and Shuman 1983	*Pediatric Pathology*	Fetal Alcohol Myopathy	"The disorder at the ultrastructural level in these damaged muscles from infants with the fetal alcohol syndrome is a unique constellation, warranting the term 'fetal alcohol myopathy.'"
Cate, Hedrick, and Morin 1983	*Clinical Chemistry* (letter to the editor)	Fetal Alcohol Syndrome and Lactic Acidosis	"Lactate measurement may be important to newborns with the fetal alcohol syndrome and metabolic acidosis."

| Garber 1984 | *Journal of the American Optometric Association* | Steep Corneal Curvature: A Fetal Alcohol Syndrome Landmark | "Keratometry is recommended as a routine test in evaluating suspected FAS children and appears to have value in screening the general school population for FAS." |
| Yellin 1984 | *Neuroscience and Biobehavioral Reviews* | The Study of Brain Function Impairment in Fetal Alcohol Syndrome: Some Fruitful Directions for Research | "It is concluded that FAS research efforts will benefit from the inclusion of psychophysiological studies." |

place on the social policy agenda, claimants may begin reconstructing the problem. Reconstructing a social problem requires revising the claims-makers' rhetoric.
In particular, claims-makers are likely to offer a new definition, extending the problem's domain or boundaries, and find new examples to typify just what is at issue.[60]

The process that Best describes as occurring in the public arena is also visible
in the medical literature on fetal alcohol syndrome. Doctors offered myriad
new symptoms to "define" and exemplify FAS. Processes of "diagnosis expansion" and "expertise expansion," parallel to "domain expansion," characterize
the next stage in diagnosis-building. Articles that exemplify diagnosis expansion were often published in the journals of medical specialties; they delin-
eated what the authors believed to be previously unrecognized symptoms of
FAS. Furthermore, such articles typify the process of expertise expansion by
staking a claim for the role of the authors' unique expertise within the territory
of this new medical phenomenon, in both diagnosis and research. There were
frequent references to the "exciting new field"[61] of FAS research or to "new
opportunities for research." These articles reported on individual case studies
or small, nonrandom samples. They typically asserted that a particular new
symptom could be used to diagnose FAS and that a particular medical specialty
could contribute to FAS research. Such articles expanded both the criteria
used to recognize and diagnose FAS and the range of specialists who could
claim authority to contribute to knowledge and research on the new syndrome.
Listed in Table 2 are some articles of this type and the claims they typically
made.

A 1981 item in the Medical News section of *JAMA*, for example, began
with the claim that "Tortuosity of the retinal vessels—both arterial and ve-
nous—may be a dead give-away that a young patient is suffering from fetal alcohol syndrome,"[62] and it went on to report on a study of seventeen cases of
FAS presented at a recent meeting of the American Academy of Ophthalmol-
ogy. Vessel tortuosity was a "new finding" and "may be of help in diagnosing
questionable cases of fetal alcohol syndrome," according to this article.[63]

Likewise, a 1980 article in *Teratology* claimed, "The abnormalities of
dermatoglyphics [the pattern of palmar creases and whorls] reported here constitute a valuable marker trait of the teratogenic effect of alcohol on fetal development and provide additional diagnostic signs for the fetal alcohol syndrome."[64] A 1982 article in Spine asserted that "the occurrence of neck fusion
in the fetal alcohol syndrome is common enough for it to be used in making

the diagnosis."[65] Claims about "marker traits" or symptoms that are a "dead giveaway" both enlarge the diagnosis and attempt to identify the elusive feature that distinguishes this syndrome from others. Occasionally, this zeal to find new diagnostic tools for FAS was extreme, as in a 1984 article in the *Journal of the American Optometric Association*; the authors reported that steep corneal curvature was a symptom of FAS, concluding, "Keratometry is recommended as a routine test in evaluating suspected FAS children and appears to have value in screening the general school population for FAS."[66] Presumably, the mental retardation that is a hallmark of FAS would be a clearer marker of the syndrome in schoolchildren than the shape of their corneas!

Articles of this type typically reported on noncontrolled studies, based on observations of small numbers of children affected with FAS. However, this approach is flawed in that it looks for clusters of defects only in the children of women who are known alcoholics and for additional markers of FAS only in children who are already diagnosed with FAS. There is no comparison with or reference to the distribution of these defects or diagnostic markers in the population at large. For example, what do we know about the incidence of steep corneal curvature, or retinal vessel tortuosity, or spinal fusion in the general population? Are they truly distinctive features of FAS only? Like much of the medical and scientific literature, the literature on FAS is oriented towards positive findings.

Diffusing the Diagnosis: FAS 101

The final stage of the early literature on fetal alcohol syndrome, which I have designated "FAS 101" articles, was also typically published in specialty journals. These articles did not make new diagnostic or research claims; rather, they were review articles intended to acquaint medical subspecialists with the diagnosis. They illustrate the process of educating a vast, diverse population of doctors and other medical practitioners, who are increasingly confined to the realm of their own specialty, about fetal alcohol syndrome. The FAS 101 articles tended to cite heavily the three original *Lancet* papers,[67] although they appeared as much as ten years later. The articles typically claimed that FAS is a major public health problem, or the leading cause of mental retardation, or the leading preventable cause of birth defects.[68] Members of the original Seattle team wrote or cowrote many of the FAS 101 articles, such as those by Streissguth in the *American Journal of Orthopsychiatry*[69] and *American Journal of*

Table 3. Sample of Articles Illustrating "FAS 101"

Author(s) and Date	Journal	Title of Article
Jones 1975	*Addictive Diseases: An International Journal*	The Fetal Alcohol Syndrome
Altman 1976	*Journal of Pediatric Ophthalmology*	Fetal Alcohol Syndrome
Luke 1977	*American Journal of Nursing*	Maternal Alcoholism and Fetal Alcohol Syndrome
Corrigan 1976	*Texas Medicine*	The Fetal Alcohol Syndrome
Streissguth 1977	*American Journal of Orthopsychiatry*	Maternal Drinking and the Outcome of Pregnancy: Implications for Child Mental Health
Robinson 1977	*Developmental Medicine and Child Neurology*	Fetal Alcohol Syndrome
Streissguth 1978	*American Journal of Epidemiology*	Fetal Alcohol Syndrome: An Epidemiologic Perspective
Rivard 1979	*Journal of School Health*	The Fetal Alcohol Syndrome
Chernoff 1979	*Currents in Alcoholism*	Introduction: A Teratologist's View of FAS
National Institute of Child Health and Human Development 1979	*Pediatric Annals*	Fetal Alcohol Syndrome
Smith 1979	*Hospital Practice*	The Fetal Alcohol Syndrome
Rosenlicht, Murphy, and Maloney 1979	*Oral Surgery, Oral Medicine, Oral Pathology*	Fetal Alcohol Syndrome
Abel 1980	*Psychological Bulletin*	Fetal Alcohol Syndrome: Behavioral Teratology
Toutant and Lippman 1980	*American Family Physician*	Fetal Alcohol Syndrome
Lindor, McCarthy, and McRae 1980	*Journal of Obstetric, Gynecologic, and Neonatal Nursing*	Fetal Alcohol Syndrome: A Review and Case Presentation
Beagle 1981	*Journal of the American Dietetic Association*	Fetal Alcohol Syndrome: A Review
Francis 1982	*Journal of Practical Nursing*	Fetal Alcohol Syndrome
Marbury et al. 1983	*American Journal of Public Health*	The Association of Alcohol Consumption with Outcome of Pregnancy
Bader and Weitzman 1983	*American Journal of Optometry and Physiological Optics*	Fetal Alcohol Syndrome

continued

Table 3. Continued

Author(s) and Date	Journal	Title of Article
Stanage, Gregg, and Massa 1983	*South Dakota Journal of Medicine*	Fetal Alcohol Syndrome—Intrauterine Child Abuse
McCarthy 1983	*Nurse Practitioner: American Journal of Primary Health Care*	Fetal Alcohol Syndrome and Other Alcohol-Related Birth Defects
LeFrancois 1984	*Vermont Registered Nurse*	Fetal Alcohol Syndrome, Maternal Alcohol Ingestion: Serious Consequences for the Fetus
Ouellette 1984	*Journal of Dentistry for Children*	The Fetal Alcohol Syndrome
Davis and Frost 1984	*Journal of Community Health Nursing*	Fetal Alcohol Syndrome: A Challenge for the Community Health Nurse

Epidemiology,[70] by Smith in *Hospital Practice,*[71] and by Jones in *Addictive Diseases.*[72] Two general categories of journals published FAS 101 articles: regional medical association journals, such as *Texas Medicine* and *Vermont Registered Nurse*, and journals of professional associations, such as *Journal of the American Dietetic Association, American Journal of Optometry and Physiological Optics*, and *Journal of Dentistry for Children*. Some articles of this type are listed in Table 3.

The FAS 101 articles typically exemplify a second type of diagnosis expansion: an enlargement of etiology and therefore of risk. In their orientation and tone, FAS 101 articles rapidly shifted the locus of risk from *alcoholics* to alcohol *use*, especially what authors called "social drinking." Whereas the classic diagnosis outlined by the Seattle team in the *Lancet* articles stipulated that the mother be a "chronic alcoholic" and that the child manifest a specific constellation of symptoms, the diagnosis of FAS expanded over time to presume that any and all drinking during pregnancy harms the child in diffuse and multiple ways, another instance of domain expansion, or "reconstructing the problem."[73] Claims such as "FAS ranges in severity from barely perceptible signs to debilitating abnormalities,"[74] often made without proof, altered the diagnosis as originally articulated and broadened the social problem. In fact, the very terms used to denote the diagnosis proliferated to include "fetal alcohol effect," "alcohol-related birth defects," and even "possible FAS."[75] One author asserted, "As this is not an 'all-or-none' phenomenon, a milder degree of damage must occur, and this may be a contributing factor in babies who fail to thrive, children with minor motor disturbances and children with school learning problems."[76] "Failure to thrive" is just the kind of diffuse, common, highly nonspecific condition that is attributed to fetal alcohol exposure.

By the mid 1980s, drinking during pregnancy was beginning to be held accountable for a whole range of medical and psychosocial problems in children. As Smith contended, "The pertinent clinical question is not 'Does this child have FAS?' but 'Is this child's problem secondary to alcohol exposure *in utero?*'"[77] And "the implications of the existing data reach far beyond the alcoholic mother"[78] — suggesting that *any* drinking by a pregnant woman was dangerous. According to this formulation, diagnosing a child with FAS or FAE at times seems less a clinical exercise than a kind of moral crusade. Smith's reasoning places implicit blame for the medical and psychosocial problems of children on the mother's drinking during pregnancy. Thus, doctors and researchers shifted concern from chronic alcoholism to moderate or "social" drinking. The U.S. surgeon general's 1981 warning exemplifies this expansion of risk: "The Surgeon General advises women who are pregnant (or considering pregnancy) not to drink alcoholic beverages and to be aware of the alcoholic content of foods and drugs."[79] Note that it targets not just alcoholic women but all pregnant and *potentially* pregnant women. Moreover, it suggests that even minuscule amounts of alcohol may be teratogenic. Fetal alcohol syndrome thus enjoys "the rhetorical advantages of broad definitions."[80] The scope of the problem depends on the definition of the syndrome or the range of congenital and developmental problems in children attributed to alcohol exposure in utero. The broader that range, the wider is the social problem represented by FAS and the greater the medical, social, and moral imperative to do something about it.

Diagnosis expansion also made possible the portrayal of drinking during pregnancy as a widespread public health problem, through such claims as "FAS is the leading cause of mental retardation" or the leading cause of "preventable birth defects." Because the emphasis on the harm associated with drinking during pregnancy shifted from women who were alcoholics to women who drank alcohol in any amount, the numbers of women and children "at risk" of harm from prenatal drinking grew enormously. Fetal alcohol syndrome was no longer a concentrated social problem but a diffuse one.

This shift in the boundaries of the problem of drinking during pregnancy — from a problem among "chronic alcoholics" to one of "social drinking" — is readily apparent in the changes in coverage of alcohol and pregnancy in obstetrics textbooks over a roughly twenty-year period, from the early 1970s through the end of the 1980s. The evolution of knowledge about alcohol and pregnancy in *Williams Obstetrics*, first published in 1903 and one of the oldest and most widely used obstetrics textbooks in American medicine, is typical. The twelfth edition, published in 1961, contains but a single mention of alcohol: "the use of alcohol by the pregnant woman has not been shown to produce any pathologic changes in mother or fetus, and no effect on the course of labor or pregnancy has been demonstrated"[81] — foreshadowing Montagu's similar claim a few years later.

After 1970, the references to alcohol in *Williams* begin to proliferate. Even before the seminal *Lancet* articles, the fourteenth edition of *Williams* (1971)[82] discussed two adverse consequences of alcohol use in pregnancy: the occurrence of delirium tremens in the newborn and the possibility that "maternal chronic alcoholism may lead to fetal underdevelopment." Despite these sequelae, this edition still maintained that "although alcohol readily crosses the placenta, its use in moderation has not been shown to produce pathologic changes in the mother or fetus or to affect the course of the pregnancy."[83]

Subsequent editions of *Williams* contained ever lengthier and fuller discussions of alcohol in pregnancy. The fifteenth edition (1976)[84] continued to assert that moderate alcohol use does not cause pathology in the pregnant woman or the fetus. However, it also suggested that "chronic alcoholism may lead to fetal maldevelopment," citing Jones, Smith, and associates and noting that "they recommend that serious consideration be given to early pregnancy termination in alcoholic women."[85] *Williams*'s treatment of the question of alcohol and pregnancy took on a very different tone in the sixteenth edition (1980).[86] The text no longer mentioned moderate drinking at all; rather, it claimed that "Excessive ingestion of alcohol by the expectant mother can produce abnormal changes in the fetus" (320). It still included a discussion of delirium tremens and other sequelae of acute intoxication and subsequent withdrawal in the newborn. The discussion of "fetal maldevelopment" was expanded, and this edition used the term *fetal alcohol syndrome* for the first time. It also expanded its recommendations on pregnancy and alcoholic women:

> Women with chronic and severe drinking problems should be discouraged from becoming pregnant until these problems are brought under control. Serious consideration should be given to early termination in alcoholic women.

> Hopefully, the adverse effects of alcohol on the pregnancy do not persist after the woman stops drinking. However, a child with most of the stigmas of the fetal alcohol syndrome has been born to parents who drank heavily in the past but denied any consumption of alcohol while members of Alcoholics Anonymous for 1½ years. (321)

The last case mentioned above engendered a sustained debate in the medical literature for several years before it was finally settled that alcohol had teratogenic but not mutagenic properties. Oddly, the sixteenth edition of *Williams* was also the first to include a lengthy discussion of ethanol as a tocolytic agent. It suggested a very specific regimen of ethanol infusion and noted what it delicately called the "troublesome side effects in both the mother and the fetus-infant" (933). This edition also included the following precaution: "Since ethanol has been widely used and acclaimed to effectively arrest labor, it is not surprising that some obstetricians have encouraged its long-term consumption prophylactically by women considered to be at high risk of going into labor before term. Ethanol should not be so employed because of the likelihood of serious effects on the fetus from long-term use" (934). Certainly this warning suggests that obstetricians commonly recommended alcohol for certain patients.

Another major obstetrics textbook with a more recent origin, *Danforth*, follows a similar progression. The first edition, published in 1966, contained no reference whatsoever to alcohol. The second edition (1971), cautiously noted that "Ethyl alcohol, in the form of cognac cocktails or dilute intravenous solution, is likewise under investigation for allaying premature labor because of its alleged inhibitory action against the release of posterior pituitary secretions, both antidiuretic and oxytocic; but adequately controlled studies are yet to be published."[87] The reference to cognac cocktails suggests again that physicians may have recommended that certain patients drink alcohol in order to reduce the likelihood of preterm labor.

The third edition of *Danforth* (1977) claimed that "intravenous alcohol is the most extensively used agent for the attempt to stop premature labor" and that "there is general agreement that it can indeed arrest certain cases of early labor."[88] Whether this treatment was effective at stopping or delaying preterm labor was a subject of much debate then and now. The factors that precipitate preterm labor remain one of the greatest obstetrical mysteries.[89] One of the respondents in my interviews recalled, "It didn't work. Then again, 80 percent of people we think are in preterm labor really aren't anyway, so 80 percent of the time it did really good." The fourth edition of *Danforth* (1982) continued to claim that ethanol was effective in stopping labor in certain patients, although it provided no guidelines about who those patients might be and it included the caution that "when alcohol is given intravenously, inebriation and belligerence may be extremely troublesome and may also complicate the choice of anesthetic if it should be needed."[90] However, this edition also included a fairly lengthy description of FAS, beginning with the observation that "like many other agents and practices long thought to be harmless in pregnancy, alcohol has now surfaced as deleterious."[91]

As "agents and practices long thought to be harmless" were reconstructed as dangerous, the circle of risk both encompassing pregnancy and circumscribing and delimiting it expanded. Thus, the fourth edition of *Danforth* contended that "total abstinence [from alcohol] appears to be the best policy in pregnancy."[92] As what was considered risky in pregnancy widened, what was considered permissible contracted. The fifth edition (1986) recommended "Alcohol intake should be sharply reduced — or, preferably, eliminated — during pregnancy."[93] This edition also included a section "Social Habits That May Adversely Affect Pregnancy," which claimed that FAS was "rediscovered in the early 1970s."[94] Finally, by 1986, the use of intravenous ethanol as a tocolytic had fallen out of favor, according to *Danforth*: "The questions of its contribution, if any, to the fetal alcohol syndrome and its effect on the neonate's respiratory and circulatory adaptation are still unsettled. Because of these questions, plus the drug's extremely unpleasant maternal side effects, the use of ethanol as a tocolytic agent has been abandoned."[95]

For a period in the 1970s and even in the early 1980s, it was not uncommon for obstetrics books to mention both the harmful effects of alcohol consumption during pregnancy and the potential benefits of alcohol as a tocolytic to arrest preterm labor. Occasionally, babies born to mothers who underwent this extremely unpleasant treatment were born with "signs of alcohol intoxication, hypoglycemia, or acidosis."[96] Interestingly, one textbook sought to make a connection between intravenous ethanol and FAS, claiming that "long-term use [of this therapy] may incur a risk of inducing the fetal alcohol syndrome."[97] However, most books mentioned both FAS and alcohol as a tocolytic without drawing any connection between the two.

"Historical Evidence" and the Rhetoric of Rediscovery

The claim in obstetrics textbooks that fetal alcohol syndrome was "rediscovered" in the 1970s echoes a broader phenomenon in the medical literature on the syndrome. The report of an association between maternal alcoholism and birth defects was initially treated as a new observation. "This seems to be the first reported association between maternal alcoholism and aberrant morphogenesis in the offspring," wrote Jones, Smith, and coauthors in their first *Lancet* article.[98] "Past evidence from animal experiments *and human experience* has not given clear indication of an association between maternal alcoholism and aberrant morphogenesis in the offspring" (emphasis added).[99] Yet only six months later, Jones and Smith published an article claiming, "Historical reports indicate that the observation of an adverse effect on the fetus of chronic maternal alcoholism is not new."[100] A rhetoric of rediscovery thus entered the early medical literature on FAS. This rhetoric and its strong moral undertones are one of the most salient features of the early literature on FAS. Claims that an association was "recognized at least 250 years ago,"[101] has been "suspected for centuries,"[102] and even observed "since antiquity"[103] permeate the literature. Authors claimed that FAS was "not a new observation"[104] or that "it is not new but has been rediscovered several times."[105] Although concerns about alcohol's effect on reproduction may have been re-ignited or even "rediscovered," in the language of Jones and Smith, in the 1970s, no rediscovery of fetal alcohol syndrome was possible, since FAS was an entirely new entity in the medical imagination. Moreover, the rhetoric of rediscovery in the modern literature on FAS draws but rarely on the medical literature of the preceding century, relying instead on a different body of "historical evidence." However, the "evidence" used to bolster these claims of rediscovery hardly meets acceptable modern standards of medical evidence; indeed, it often originates in antiquity.

For example, one of the most common historical references in the early medical literature on FAS was to an ancient Greek and Roman belief that intoxication at the time of procreation results in the birth of a damaged child, as expressed, many authors claimed, in the myth of Vulcan.[106] Some articles even noted that "Vulcan, the deformed blacksmith of the gods, was said to be the result of such influence."[107] Several articles also referred to passages from Greek "philosopher/scientists," as they were called by one author.[108] Aristotle's contention that "Foolish, drunken and harebrained women most often bring forth children like unto themselves, morose and languid" was quoted by an article in *Pathology Annual*.[109] An article in *Human Biology* cited a passage from Plato's *Laws* that "It is not right that procreation should be the work of bodies dissolved by excess wine, but rather that the embryo should be compacted firmly, steadily and quietly in the womb."[110]

Another common reference was to an ancient Carthaginian custom prohibiting drinking on the wedding night. This custom was attributed to a concern to prevent conception under the influence of alcohol.[111] The Judeo-Christian tradition also yielded purported evidence in the form of an Old Testament verse (Judges 13:7) in which an angel speaks to a woman, telling her "Thou shalt conceive, and bear a son; and (therefore) now drink no wine or strong drink, and eat not any unclean thing."[112] This verse even stood as an epigram to a 1978 article on FAS by Sterling Clarren and David Smith in the *New England Journal of Medicine*.[113]

All these quasi-historical allusions were misconstrued. "Nearly all the statements investing the ancient and medieval past with precognition of this disorder are wrong."[114] In ancient Greece, Rome, and Carthage, the belief was not that drinking during pregnancy harmed the child but that intoxication at *the moment of conception* — not during gestation — led to deformity. The ancient Romans also used wine as an abortifacient,[115] another reason that pregnant women may have been warned not to drink. Furthermore, the ancient belief was that the *father's* drunkenness could affect conception, as much as or more than the woman's. Authors' frequent references to the lame god Vulcan are particularly curious; in ancient myth his deformity is never attributed to his mother's drunkenness — in fact, his lameness is not even a congenital defect, but the result of injury.[116] The Old Testament passage was misinterpreted as well. The injunction to abstain from drink stems not from any awareness of the dangers of drinking during pregnancy but from the child's status as a Nazirite; the Nazirites were a group forbidden, among other things (including cutting their hair), to take any intoxicants.[117] Moreover, as Mary Douglas notes, ritual prohibitions are just that: ritual and symbolic as well as arbitrary and thus not necessarily based on evidence or experience of harm from exposure.[118]

In addition to these ancient references, two studies from the nineteenth century were also singled out in the early literature on fetal alcohol syndrome. The first was the 1834 report to the House of Commons in Britain, noting the "starved, shrivelled and imperfect look" of children born to heavy drinkers.[119] The second was Sullivan's 1899 study of women in a Liverpool prison, which found that among six hundred children of 120 women, spontaneous abortion and stillbirth rates were higher for chronic alcoholics and that their surviving children had an increased frequency of epilepsy.[120] However, only these two studies were repeatedly cited in the early medical literature on FAS, suggesting that modern physicians were largely unaware of the earlier literature on alcohol and reproduction.

One of the most interesting pieces of historical "evidence" marshaled in support of FAS was a 1751 engraving by William Hogarth (Figure 4). Made during England's Gin Epidemic, the print depicts the chaos and social disorder attributed to the sudden abundance of cheap gin among London's lower classes. Although the engraving was meant to be social satire, an exaggeration rather than an accurate portrayal, a 1981 article in *JAMA* treated the print quite literally and attempted to diagnose FAS in one of the figures depicted in it. The author, Alvin Rodin, wrote, "The facial appearance of the child (falling off the lap of the woman seated in the foreground) is worthy of note. The eyes have a shorter than normal palpebral fissure, resulting in relatively round, 'Orphan Annie' eyes. These are one of the features of the effects of maternal alcohol on fetal morphogenesis, described by Jones et al. as part of the fetal alcohol syndrome."[121] However, many of the figures in the print are depicted with unusually round eyes; Rodin mistook a stylistic device on Hogarth's part for medical evidence. In fact, in the companion piece to *Gin Lane*, called *Beer Alley*, in which Hogarth contrasts the salubrious nature of beer drinking to the debauchery induced by gin, many of the figures have the same round eyes. A review article in the *Annals of the New York Academy of Sciences* also used Hogarth's engraving, claiming that "the relationship of intoxication to child abuse and neglect is apparent."[122]

The rhetoric of rediscovery took on a life of its own; any historical mention of alcohol, pregnancy, and birth defects was inflated to evidence. It is telling that authors often reported the particulars of this "evidence" incorrectly. For example, the Carthaginian prohibition against drinking on the wedding night was at times referred to as a Greek or Roman custom.[123] The date of the 1834 report to the House of Commons was variously cited as "in the 1700s,"[124] in the "eighteenth century"[125] and in 1934.[126] One article dated Sullivan's 1899 study as 1800.[127] Since historical evidence was repeated uncritically from one article to the next, these errors proliferated in the literature. The imprecision of these references, atypical in the highly detail-conscious medical profession, betrays the latent functions of this so-called evidence.

FIGURE 4. William Hogarth's *Gin Lane* (1751). Graphic Arts Collection, Department of Rare Books and Special Collections, Princeton University Library.

Doctors thus invented a history of fetal alcohol syndrome. Indeed, such "historical evidence" is unusual in the medical literature and demands that we ask why it is brought to bear on FAS. What is the purpose of these citations to passages from the Bible, from myth, from ancient texts, from art history, and from nineteenth-century England — "the alleged history" of FAS?[128] The rhetoric of rediscovery took on such an important role in the early literature on FAS for four main reasons.

First, these historical and quasi-historical references were an attempt to augment the sparse data initially available to document the existence of the newly recognized syndrome. Faced with extremely small numbers of cases, FAS entrepreneurs sought to bolster the available evidence with references to the past. Early articles on FAS proceeded almost formulaically, marshaling ancient evidence, then moving forward through time to Jones and Smith.[129] One author publishing in *JAMA* in 1981 went so far as to put the historical record on a par with contemporary evidence, claiming, "At present we are only a little more advanced in our knowledge of the pathogenesis of the fetal alcohol syndrome; no more concise statement of the overall problem has been made than that of the Middlesex Sessions Committee in 1735: 'Unhappy mothers habituate themselves to these distilled liquors, whose children are born weak and sickly, and often looked shrivel'd and old.'"[130] By claiming that knowledge has not advanced since the eighteenth century, this physician privileges historical reports, suggesting that in the case of fetal alcohol syndrome, we need not look to scientific or even contemporary evidence to find proof of the authenticity of the diagnosis. History and myth are advanced as acceptable — perhaps in this case preferred — ways of identifying and verifying medical conditions.

The second function of these historical references was their potential to refute the dearth of present-day experiential evidence of any correlation between drinking during pregnancy and poor birth outcome. Medical observers who sought to locate FAS in the late twentieth century were faced with the task of explaining why, when women have been drinking and giving birth for thousands of years — or as one skeptical doctor noted, when "alcohol ingestion has been a feature of the human diet at least since the invention of glazed pottery . . . and the association of oral ethanol intake with subsequent illness of the imbiber is so well-known as to be cliche"[131] — an association between drinking and birth defects had never been, in their collective memory, commonly acknowledged. Despite the claim of an article in *Pediatric Annals* that the association between drinking and poor birth outcome was "conventional wisdom,"[132] the early FAS entrepreneurs believed there was very little to suggest that doctors or laypeople had noted such an association, much less drawn firm conclusions about drinking during pregnancy and birth outcomes.[133] Even today, thirty years after the diagnosis was introduced, many doctors still express skepticism about the relationship between moderate or social drinking and birth defects. An older family practitioner mused,

Certainly when I was pregnant, I drank. I didn't smoke, 'cause I didn't know how to at that point. You know, I think of all the kids that are grown up in our generation that don't seem to have these physical attributes [of FAS] and learning disabilities and I wonder why not because we all drank like, not like fish exactly. But certainly I would never give that advice to anybody now. I think that people are much more generally conscious that anything that goes into your body pretty well gets through to the fetus and we don't know the full effect of that. [29F, female]

This doctor's remarks illustrate how drastically medical and lay conceptions of pregnancy and risk have changed over the course of her adult life. From drinking blithely, almost "like fish," during pregnancy, to being conscious that the fetus is exposed to "anything that goes into your body," laypeople's and physicians' knowledge underwent a complete transformation and radical reversal. Nevertheless, even as she acknowledges this shift, this doctor questions it in the light of her own experience. Later in the interview she recalled that "back in the 50s and 60s, women were drinking every night at cocktail parties. We were all drinking. We all should have been affected, but we weren't, so why is that?"

Similarly, an obstetrician who began practicing medicine in the mid-1960s told a story that illustrates how his own observations seemed more powerful than the official medical knowledge. Interestingly, as a resident this doctor had known Kenneth Lyon Jones, one of the authors of the original *Lancet* articles on FAS, and had discussed the diagnosis with him before it officially entered the medical literature. Nevertheless, he weighted his own observations more heavily in constructing risk during pregnancy and formulating advice for his patients.

I have to give you just a little anecdote that certainly colored the way — my advice. I spent a summer in France when I was an exchange student . . . The family that I lived with, the brother had a sister who was married and had four children . . . I saw their dietary habits over the course of a two-week period of time that we stayed with them . . . Everybody each day had wine. Dominique, the mother, had wine . . . Certainly it was cut wine, I mean it was not sort of solid or 100 percent wine, it was diluted wine. That was their lifestyle. She did not, and I am sure she — I don't believe that she drank just water since they don't drink just water, during her pregnancy. So as I thought about it with my own patients and the advice I gave them, I harked back to my experience in France with wine, thinking that probably Dominique drank wine on a daily basis with each one of her four pregnancies and her children were just as normal as could be. [25B, male]

Some doctors mentioned that "our parents used to drink" and often at levels that today would be considered risky. "To be honest, I am sure that lots and lots of our parents drank during our whole pregnancy, their entire pregnancy with us, a beer or two, or glass of wine or two, and most of us are fine" [26B, male]. In other words, doctors' intuitive or experiential knowledge often contradicts official constructions of drinking during pregnancy as risky. Moreover, thinking of alcohol use as absolutely contraindicated during pregnancy went against the grain not only of doctors' social experience and observations but of their clinical acumen as well. "Well, it's interesting that there never was such a thing as fetal alcohol syndrome until the early 70s," mused one elderly pediatrician [10P, male].

Thus, observers of FAS in the late twentieth century cast their nets far back into the past, creating a history of FAS that relied on questionable "evidence," in order to persuade their audience that prior human experience *had* noted the ill effects of drinking during pregnancy. Yet there are alternative explanations for this dearth of prior evidence. Heavy alcohol consumption raises the risk of spontaneous abortion, and under the conditions of high fetal and infant mortality prevailing in the past, fewer alcohol-affected babies may have survived birth or infancy. Moreover, the rate of birth defects overall was probably higher in the past, and FAS may not have been distinguishable among them. And drinking among women may have been so common (as surely must have been true for certain populations at certain times) that no independent association between drinking and birth defects could be easily observed.

Interviewed physicians had their own explanations for why FAS was not discovered until 1973. "Well, because the stigmata, although quite obvious when you know what they are, are relatively subtle and could very well be attributed to just this being a funny looking kid. And there are lots of causes of mental retardation, and you never know in the great majority of them where it is coming from, so this didn't occur to us, I think, to ask, 'Were you drinking a lot during pregnancy?'" [10P, male]. Notice that according to this explanation, there is no reason that FAS ever should have been discovered. The stigmata remain subtle; the causes of mental retardation are still multiple and not well understood.

Some doctors offered professional advancement as another explanation for why the diagnosis of fetal alcohol syndrome took hold in medicine after it was discovered. "There's a badge of honor in having [i.e., diagnosing] some rare disease, because all you see is all this common stuff," said one pediatrician [6P, male]. Catching the rare diagnosis of FAS signals a doctor's clinical acumen and savvy. An obstetrician who specialized in maternal-fetal medicine thought that "people are trying to make professorships" out of this debate about FAS and FAE [4B, male].

Doctors also speculated that, ironically, the very uncertainty of the diagnosis and the disorder it represents contributed to the establishment of the definitive diagnosis. One pediatrician said, "I used to know every syndrome and then you start to see that it's all man-made . . . you realize it's not a science" [6P, male]. To this doctor, the taxonomy of birth defects became less important as he became older and more experienced. "So you get a little jaded and it becomes less important and less interesting because there are so many other things that take your time." He also thought that some of the impulse to create diagnostic categories for new syndromes comes from a feeling of medical impotence. "And then we go into things like [there's] nothing to do about it directly, so you name it." A family practitioner likewise spoke of the futility of making the diagnosis: "So I don't think making the diagnosis actually helps. More than just, like, not exploring their funny-looking kid status anymore, you know, if they are a funny-looking kid and you want to know why, and you can't find a reason and you suspect FAS and you find some confirmation for that. They have the perfect characteristics and a history of mom drinking, then I would say, the only reason to make the diagnosis of FAS is to say we don't need to look further . . . The diagnosis, I don't think, implies any value" [30F, male]. Doctors also acknowledged that patients themselves often want diagnoses, official descriptions of their sufferings, so as to validate the condition. One family practitioner referred to diagnoses of ambiguous or contested disorders like fetal alcohol syndrome, chronic fatigue syndrome, and Lyme disease as "security blankets" for patients and physicians alike. Formal diagnosis reassures in a world of uncertainty, ambiguity, and unmanageability. Naming the disease, even if the physician can do nothing else, eases the patient's suffering.

The third function of historical references is that the invented history of FAS allowed writers to illustrate what they believed to be the timelessness and persistence of drinking during pregnancy as a human problem. Cries that "we cannot afford to repeat history"[134] were typical. An article in the *Journal of Obstetric, Gynecologic, and Neonatal Nursing* lamented, "It is both appalling and unfortunate that these warnings were not heeded until recent times."[135] Historical references thus enabled modern observers to express their own indignation and to register the gravity of the situation. "Claims need to be compelling if they are to be successful."[136] Claims about a forgotten or lost history dramatized the new syndrome.

Finally, the moral judgment implicit in most of the so-called historical evidence marshaled in the early literature on FAS served to exaggerate further the significance of the new syndrome. There is an undeniable element of moral condemnation in many of the historical references. This censure is most apparent when the Bible is quoted. Several articles enjoined that "until much more is known, it would seem advisable to observe the Biblical admonition that 'thou shalt conceive, and bear a son; and (therefore) now drink no wine or strong drink' (Judges 13:7)."[137] Other authors offered this verse as an epigram. Whenever doctors urge women not just to follow their advice but to heed the Bible, they invoke moral as well as medical authority. The use of Hogarth's engraving also carried a moral weight of condemnation: it depicts a scene of chaos, of extreme disregard for human life, and of abandonment of maternal responsibility. Its appearance in the pages of *JAMA* in an article on FAS projected Hogarth's rebuke of extreme drunkenness during England's Gin Epidemic onto contemporary American women who drink during pregnancy, more than two centuries and an ocean distant from the target of Hogarth's social commentary.

The Language of Opprobrium

The very language used in much of the early medical literature on fetal alcohol syndrome also betrays the moral orientation of the writers. Scientific writing often evinces literary rhetoric.[138] In the medical literature on FAS, both word choice and metaphor frequently illustrated opprobrium, even when couched within ostensibly neutral medical terminology. Although the term used to denote the diagnosis in the second article by the Seattle team[139] has persisted, other writers have proposed and used such terms as "alcohol embryopathy"[140] and "fetal alcoholic syndrome,"[141] and even "fetal alcohol abuse syndrome,"[142] which doubles the moral censure by implying that not only does the woman abuse alcohol, she also abuses her unborn child. The term *fetal alcoholic syndrome* seems to revert to nineteenth-century notions of heredity — as if the baby could be born a drunkard by virtue of the mother's intemperance, a direct transmission of moral lassitude. It is interesting to speculate on why the syndrome was not named eponymously, as are so many patterns of malformation — and diseases more generally.[143] Perhaps "Jones-Smith syndrome" simply sounded too quotidian, as I heard one doctor suggest at a medical conference; however, it is more likely that fetal alcohol syndrome was so named to highlight what its discoverers believed to be its cause. The proliferation of other labels containing the word *alcohol* supports this supposition.

The meaning of both the term and the diagnosis has widened over time to include "possible fetal alcohol syndrome," "partial FAS," "fetal alcohol effect," "alcohol-related birth defects," and "alcohol-related neurodevelopmental delay."[144] Note that this widening encompasses a larger role for alcohol in the etiology of adverse birth outcome but does not broaden the possible causes or explanations for what is observed and diagnosed as FAS. The author of a letter to the editor of the *New England Journal of Medicine* who wryly noted that "alcohol abuse may have a role in the 'fetal alcohol syndrome,' but the syndrome may eventually prove to be a 'polydrug-abuse syndrome' or conceivably 'a polydrug-abuse-nutritional deficit-stress induced fetal syndrome'"[145] was a rare voice of caution indeed. Far more common were such reproving titles as "The Tragedy of Fetal Alcohol Syndrome,"[146] "Fetal Alcohol Syndrome — The Incurable Hangover,"[147] and "Fetal Alcohol Syndrome — Intrauterine Child Abuse."[148] Just as in the social construction of the drinking-driving debate, in the discourse about fetal alcohol syndrome "the rapidity with which alcohol is perceived as villain exemplifies the moral character of factual construction."[149]

Not surprisingly, women, their actions and their bodies, were often referred to in harsh terms reminiscent of late-nineteenth-century discourse about women who "poisoned" their fetuses: "maternal alcoholic environment,"[150] "recidivist maternal alcoholics"[151] (echoing the language of criminology), "fetal hazard,"[152] "acute fetal poisoning,"[153] "fetal damage,"[154] and "embryo-toxin"[155] were all used to describe women and their bodies. Although in some cases these terms are standard medical jargon, they nonetheless bear metaphorical force. Writers typically focused on the "harsh intrauterine environment"[156] created by women who drink during pregnancy: in the eyes of these writers, such women have clearly failed to fulfill their roles as nurturers. One writer commented bluntly, "The intrauterine environment of the alcoholic mother is a risky and complex milieu within which her fetus develops and grows. She inflicts upon her fetus the potential teratogenic effects of not only alcohol but also often adds the insults of malnutrition, hepatic cirrhosis, infection, smoking, drug abuse, hyperpyrexia (fever), and trauma."[157] Here, the effect of the woman's own poor health on the fetus is made to seem intentional, almost willful: she "inflicts upon the fetus" and "adds the insults," as though she deliberately chooses to be malnourished, sick, addicted, and traumatized and thus to undernourish, infect, addict, and traumatize her fetus. This is the language of combat and aggression, not maternal nurturing. Moreover, describing the pregnant woman's uterus as an "environment" draws on the metaphor of physical, geographical space to transform the woman's body itself into an inert landscape, a space or "environment" that the fetus inhabits, rather than an intimate, life-sustaining connection between the woman's body and the fetus. (Referring to the pregnant woman's uterus as the "maternal environment" is a rhetorical tactic of the famous antiabortion film *The Silent Scream* as well.)

Such language also reveals the use of FAS as a stigmatizing condition. Indeed, the literature contains several examples of how FAS was and is used to label affected women and children as medical and social problems. Most commonly, the diagnosis of FAS in a child was used to diagnose the mother as an alcoholic retrospectively.[158] As Smith noted, "Clinicians familiar with FAS have been alerted to previously unrecognized alcoholism in a mother simply because her child had a distinctively 'FAS' face."[159] Erb and Andresen likewise noted in *Clinical Pediatrics* that "many children will be identified as displaying FAS and their mothers can be identified as alcohol abusers."[160] Since alcoholism in the mother was one of the original criteria for a diagnosis of FAS, this tautology seems particularly insidious, reasoning as it does that since the child has FAS, the mother is an alcoholic; therefore, the child has FAS. Indeed, the power of FAS as a morally stigmatizing category helps to explain the rapid embrace of this new diagnosis in the medical and lay world alike.

"Rhetoric is central, not peripheral to claims-making. Claims-makers intend to persuade, and they try to make their claims as persuasive as possible."[161] One way FAS entrepreneurs dramatized their claims was to use case studies as evidence. This rhetorical device is evident as well in the recent Institute of Medicine report on FAS, in which an entire chapter is devoted to vignettes of children and adults with FAS and women who drank during pregnancy.[162] "Examples of atrocities" are used to "typify social problems."[163] Case reports such as the ones included in the original *Lancet* articles are standard in medicine; they constitute an important way of knowing.[164] "The individual case is the touchstone of knowledge in medicine."[165] Yet case studies can also serve a dual purpose: "Atrocity tales do not merely attract attention; as typifications, they also shape perceptions of the problem."[166] The particular case not only is the foundation of individual doctors' knowledge of FAS but also comes to represent the diagnosis and the wider social problem.

Often, however, the humanitarian impulses of the clinicians faced with individual cases of FAS were at odds with their inclination to censure what they saw as inappropriate social behavior. The importance of diagnosing FAS in infancy was often stressed, yet just as often a note of deep pessimism was sounded. Initially, physicians and obstetrics textbooks recommended therapeutic abortion for fetuses exposed to alcohol during pregnancy — a stark dramatization of the prevailing belief that such children were irredeemably damaged. The early literature on FAS typically noted that there was little evidence of catch-up growth in FAS children or of any ameliorative effect of a positive environment or special educational intervention. Jones and Smith wrote of the "overwhelming magnitude of the handicapping problems that maternal alcohol can impose on a developing fetus."[167] The damage at birth was referred to as irreversible, and children with FAS were dismissed as beyond the realm of medical intervention. "Even the presence of a few stigmata should be considered a warning of future developmental delay."[168] The grim prognosis of Jones and coauthors in 1974 — "Thus the offspring of chronic alcoholic women, whose development and function are often permanently damaged by their adverse intrauterine environment, frequently become a problem for society in postnatal life"[169] — neatly summarizes the attitude of most of the early writers on fetal alcohol syndrome. Ironically, this prognosis also closely echoes the kind of assessment made by many of the nineteenth-century medical writers and temperance crusaders. No less in 1974 than in the nineteenth century, both the woman who drinks during pregnancy and her child were believed to be beyond hope and destined to be social problems.

CHAPTER 4

CHARTING UNCERTAINTY

THROUGH DOCTORS' LENSES

It's just one of many tragedies I see every day.
—An obstetrician-gynecologist

The medical literature explored in Chapter 3 reflects the official knowledge and construction of the diagnosis of fetal alcohol syndrome. Communications within the field, such as case reports and letters to the editors of medical journals, are one way of constructing medical knowledge and conceptions of risk. Research and review articles on FAS are another important element of the social construction of the diagnosis. These forms of published literature are one type of "doctor talk" about fetal alcohol syndrome. However, equally important is the way doctors talk about and construct the syndrome in their offices and examining rooms, in their interactions with their professional colleagues and their encounters with their patients. Medical knowledge is manifest not only in medical journals and textbooks but in the collective and individual consciousness, minds, experiences, and practices of doctors. How do doctors experience and construct the diagnosis of FAS in the everyday world of patient care, of examination rooms and clinics, delivery suites and nurseries? How do the doctors who treat pregnant women, infants, and children "talk about" the diagnosis? This chapter explores contemporary physicians' knowledge and attitudes about drinking during pregnancy, fetal alcohol syndrome, and the women and children whose lives this syndrome affects. It portrays "the sum total of 'what everybody knows' about a social world" — that of doctors — "an assemblage of maxims, morals, proverbial nuggets of wisdom, values and beliefs, myths and so forth."[1] It is from this set of everyday assumptions and actions that the social reality of the diagnosis of FAS is constructed among the medical professionals who care for women and infants.

This chapter describes how doctors learn about FAS and whether and how often they encounter it in their clinical practice. It seeks to uncover how doctors describe the diagnosis, whether they see it as "real," as "serious." It seeks doctors' evaluations of whether the diagnosis is useful to them as clinicians, as well as to their patients, be they pregnant women, mothers, or infants and children. More generally, this chapter seeks to delineate how doctors think about the risks associated with drinking during pregnancy. It describes doctors' reactions to that risk as it is embodied in their patients who drink and in their patients who suffer from fetal alcohol syndrome.

Human reproduction is an inherently risky endeavor. Despite significant advances in knowledge and technology, modern medicine is not able to forestall or even foresee many kinds of adverse birth outcomes. How, then, do doctors think about this irreducible uncertainty during pregnancy? Uncertainty presents both clinical and philosophical dilemmas. It complicates not only how doctors do their work but how they think about it as well. However, uncertainty in pregnancy is unique in that it involves not one patient but potentially two — the pregnant woman and the child she will deliver. Moreover, the uncertainty and risk inherent in reproduction manifest both an individual dimension and a social one. At stake in every pregnancy is the new life that is being brought into the world, into society. Thus, the uncertainty of pregnancy encompasses not just fears and threats to the woman's health but also questions about the health and vitality of her offspring. Thus, as a society, we feel we have a deep investment in other people's pregnancies.[2] The risks associated with a pregnant woman's behaviors, such as smoking, drinking, diet, or sexual behavior, may affect not only her own health but that of the next generation as well. How do doctors answer the question of whose welfare is at stake? The actions of the pregnant woman may be seen to engender risk for the maternal-fetal unit, the fetus, or the woman herself, depending on the perspective adopted. Given the duality of risk in pregnancy, how do doctors conceive, categorize, and classify prenatal risk associated with alcohol? Moreover, how do physicians think about responsibility — their own and their patients' — during pregnancy? The doctors' comments presented in this chapter reveal much about how social and individual responsibility is assigned and how ideas about responsibility are bound up in ideas about causation.

I describe here doctors' reports of their interactions with patients; I did not directly observe doctors counseling patients about drinking or reacting to patients with the diagnosis of FAS. Rather, I gathered their reflections and recollections about such experiences through interviews, inquiring about their knowledge of fetal alcohol syndrome and their attitudes towards the diagnosis and towards prenatal drinking compared with other kinds of risk-taking during pregnancy. I conducted interviews with obstetrician-gynecologists, pediatricians, and family practitioners.[3] These doctors practice at five teaching hospitals in Philadelphia, all serving a wide range of patients. Some draw patients from around the region, if not the country. From the doctors' reports, I have attempted to recreate their vision of FAS, to see this diagnosis and these patients through physicians' eyes.

Although we can learn much from these interviews, they do have some limitations. The sample of respondents is drawn primarily from the world of academic medicine, and many of the doctors do research in addition to seeing patients. This fact has several implications for my findings. Academic doctors may be more "knowledgeable"; that is, they may make a more concerted effort to keep abreast of medical developments. In addition, their location in academic centers may expose them more routinely to emerging knowledge. In any case, their knowledge base may not be comparable to that of nonacademic peers in medicine, who may rely less on the emerging literature and more on their own clinical experience in developing their acumen and formulating clinical decisions.[4]

These academic physicians may also have been more willing to consent to an interview, either because they shared an ethos of research or because their daily schedules afforded more flexibility than is possible for doctors whose days are filled with the time pressure of attending to one patient after another. However, I want to stress that not all the respondents in my sample were academic doctors. Many were practicing physicians whose patients were their first and foremost professional responsibility.

Most doctors in the sample reported that they saw a wide range of patients, from the wealthy and influential to the working poor and the publicly insured. However, due to the nature of their practices, some doctors saw only Medicaid patients, others saw exclusively privately insured patients. Most doctors, though, saw a range of patients and thus were able to draw on a broad base of clinical experience in reporting their interactions with patients.

Another serious limitation of my data is that almost all the respondents are white. Therefore, the data are blind to the ways in which race may influence a doctor's perception of FAS. However, I did make a concerted effort to diversify the sample along two other important demographic dimensions: gender and age. Fetal alcohol syndrome is a fairly recent addition to the diagnostic spectrum; certainly there are doctors practicing today who began practicing more than thirty years ago, before the diagnosis was created. Thus, I sought to include among my respondents doctors who had begun their careers both before and after the introduction of FAS. Seven of the doctors graduated from medical school before 1973; several others graduated and began practicing in the 1970s, before FAS was widely known. In addition, gender may be a significant characteristic shaping a doctor's attitude towards risk in pregnancy. The sample included more men (17) than women (13); however, this mix (57 percent male, 43 percent female) overrepresents women compared with the general population of doctors in the United States (79 percent male, 21 percent female).[5] (See Table 4 for an overview of respondents. Throughout this chapter, I use within brackets the codes in Table 4 to identify the speakers: *B* denotes obstetricians; *P*, pediatricians; and *F*, family practitioners.)

Within the three specialties, many doctors specialized in subfields as well. Among the obstetrician-gynecologists, four were specialists in maternal-fetal medicine. Four doctors (one pediatrician and three obstetricians) were board certified as clinical geneticists. A few of the pediatricians identified themselves as specializing in child development.[6] Doctors held a range of other specialties as well, including gastroenterology, nephrology, emergency medicine, and geriatrics. Almost all the doctors were trained in allopathic medicine; however, there were two osteopaths among my respondents.

How Doctors Learn about FAS:
"All I Know Is What I Read in That Book"

Doctors report learning about fetal alcohol syndrome through a variety of channels, which may be roughly divided into three categories: learning about FAS in medical school, as part of the curriculum, or while doing a residency or fellowship (a kind of explicit training); learning about FAS through professional experience with patients (clinical, "on-the-job" training); learning about FAS through personal experience, outside clinical work (personal knowledge). A fourth source of knowledge — witnessing FAS while sojourning as a medical volunteer, typically on an American Indian reservation[7] — is very similar to the

Table 4. Overview of Interview Respondents

I.D. No.	Primary Specialty (additional board certification)	Sex	Race	Year Graduated Medical School	Ever Seen FAS?
Obstetrics and gynecology (sex ratio: 55% M, 45% F)					
2B	Clinical genetics	F	W	1980	Y
4B	Maternal-fetal medicine	M	W	1983	Y
7B	Clinical genetics	F	W	1983	Y
14B	Maternal-fetal medicine	M	W	1990	Y, as resident
15B	Maternal-fetal medicine	M	W	1963	Y
17B		F	W	1983	N
21B		M	W	1992	Y, once
25B		M	W	1965	N
26B	Maternal-fetal medicine; clinical genetics	M	W	1972	Y
27B		F	B	1978	N
28B		F	W	1984	N
Pediatrics (sex ratio: 50% M, 50% F)					
5P	Gastroenterology	M	W	1986	Y, as consultant
6P	Nephrology	M	W	1961	Y, but not patient
8P		F	W	1993	Y
9P		F	W	1983	Y
10P	Developmental and behavioral pediatrics	M	W	1954	N
11P		M	W	1985	Y
12P	Osteopathy	F	W	1988	Y
19P	Child development	M	W	1973	Y
20P	Clinical genetics	F	W	1968	Y
24P		F	W	1985	Y
Family practice (sex ratio: 66% M, 33% F)					
1F	Preventive medicine	F	W	1976	Y, as resident
3F		M	W	1986	N
13F		M	W	1958	Y, but not patient
16F	Geriatrics	M	W	1974	Y
18F	Emergency medicine	M	W	1981	N
22F	Osteopathy	F	W	1994	N
23F		M	W	1981	N
29F		F	W	1978	N
30F		M	W	1994	N

third category: it is knowledge based on personal experience, not on clinical exposure or formal instruction.

Michael Dorris's autobiographical book *The Broken Cord*[8] represents an important variant of this personal knowledge category. Dorris was a Dartmouth professor and single father who adopted an American Indian son, and the boy was later diagnosed as having fetal alcohol syndrome. *The Broken Cord* is an angry, powerful, and moving account of Dorris's experiences raising his son, learning that the boy's severe physical and mental developmental problems are the consequence of exposure to alcohol in utero, and learning as much as he can about the disorder. He wrote of his son, "A drowning man is not separated from the lust for air by a bridge of thought—he is one with it—and my son, conceived and grown in an ethanol bath, lives each day in the act of drowning. For him there is no shore."[9] The book is part memoir, part social history of drinking and FAS among American Indians, part diatribe. The book was made into a TV movie of the same name in 1992. Both the book and the movie did much to publicize and dramatize the problem of FAS. In fact, the novelist Louise Erdrich, whom Dorris married after the adoption, wrote in the foreword, "It has been wrenching for Michael to relive this story, but in the end, I think he felt compelled to do so after realizing the scope of the problem, after receiving so many desperate and generous stories from other people, and, in the end, out of that most frail of human motives—hope. If one story of FAS could be made accessible and real, it might stop someone, somewhere, from producing another alcohol-stunted child."[10] In other words, Dorris was acting as a moral entrepreneur in writing his book.

A surprising number of doctors mentioned having heard or read about the book as their first introduction to FAS. (Only two doctors—a pediatrician and a family practitioner—said they had actually read the book.) One doctor heard Dorris speak on the radio. Doctors who had been exposed to *The Broken Cord* reported being deeply affected and impressed by it. "I think that's where I was probably more struck by what the ramifications of this could be," recalled an obstetrician [7B, female]. While the book is full of information about FAS, it is written for a lay audience and is certainly not medical in style or content. In fact, it is a highly moralized account. Yet, for a few doctors, *The Broken Cord* was their only or most vivid acquaintance with the syndrome. And all doctors who mentioned the book invested it with a high degree of authority. "All I know is what I read in that book," said one pediatrician, who in fact had spent considerable time on American Indian reservations working with children with

FAS and during the interview pulled out an article on FAS from *Social Biology* to show me. That kind of authority makes *The Broken Cord* unique among personal memoirs of illness, which are often dismissed by doctors as merely "lay accounts." Typically, doctors reject patients' constructions of new or disputed illnesses, such as chronic fatigue syndrome, fibromyalgia, and Gulf War syndrome, or even patients' constructions of medically established diagnoses like asthma or clinical depression. Perhaps because doctors so rarely encountered FAS in their own patients they were willing to invest Dorris's account with a super-ordinary authority. The significance attributed to *The Broken Cord* is also a peculiar and specific manifestation of the general predominance of the case study in medicine. Singular (and single) cases can become emblematic of an entire disease.[11] The book is also an example of an "atrocity tale,"[12] a common feature of the depiction of social problems in the public arena. "The atrocity — usually selected for its extreme nature — typifies the issue; it becomes the referent for discussions of the problem in general."[13] Dorris's son seemed to suffer from a particularly severe and disabling set of symptoms, so his case provided a singularly dramatic and compelling example of the syndrome. Dorris's description of his son's problems in *The Broken Cord* became the moral and medical personification of FAS in some doctors' minds.

A few doctors recalled reading the Jones and Smith articles that first discussed FAS in the *Lancet* in 1973. Others mentioned reading review articles in the *New England Journal of Medicine* or learning about FAS through the second edition of Smith's atlas of birth defects, *Recognizable Patterns of Human Malformation*, published in 1976.[14] One of the younger doctors in the sample, a 1994 medical school graduate, reported first hearing about FAS in health class in high school.

Not surprisingly, doctors were often prompted to learn more about FAS through necessity: when they first saw a patient with this diagnosis, or when they suspected the diagnosis, or when a patient inquired about the syndrome. "I have a whole file drawer on fetal alcohol because what I do for a living is work with kids with disabilities, so I am in a need-to-know situation, so I follow that literature. So I am probably not representative of what the typical physician would know. I hope I am not representative" [19P, male]. Occasionally, doctors were prompted to learn more about FAS or the risks of drinking during pregnancy by inquiries not from patients but from their own family members who turned to them as medical authorities.

Doctors sometimes reported clear memories of particular cases of FAS, often the first ones they saw, as important markers in their own intellectual construction of the diagnosis. For some doctors, the first case of FAS they encountered, often as a medical student or a resident, stood out vividly in their minds. One obstetrician recalled a particularly striking case:

> The other thing I remember was from a pure case standpoint. We saw a child who had what now would be considered to be classic stigmata, the whole facial features, she weighed, like, two and a half pounds at birth, and we couldn't figure out what she had and we thought it was a genetic syndrome and we asked her mother to come in and talk about it. And her mother came in and brought her five-year-old daughter with her who, again, now one realizes, had all the stigmata of FAS, all the facial features, she was small, she was really developmentally delayed, and the mother turned around to talk to us and you could just smell the alcohol on her breath . . . [The diagnosis] almost knocked us over; actually, I still have the pictures of them in my slides. [26B, male]

For this doctor, who recounts this first encounter with FAS every time he lectures on the subject and shows his slides, emotional memory plays as important a role as didactic memory in shaping clinical knowledge of the diagnosis of FAS. He remembers how he *felt* about this patient as distinctly as he recalls the clinical symptoms of the diagnosis. Moreover, as Kathryn Hunter reports in her study of the "narrative structure of medical knowledge," such anecdotes become an important link in the transmission of medical knowledge.[15] This case serves as an archetype of the diagnosis not only in the doctor's own mind but in the minds of the students and residents who see his slides. This case may have had such striking resonance for this doctor because of the patient's history, as he subsequently learned it. "It was an interesting story when we finally sat down with her [the mother] and she told us the whole story. When she was pregnant with her second child, she actually used to take the first child with her to the bar and from, like, eight in the morning till, like, three at night she would sit on one barstool and her other daughter would sit on the other barstool and at the time she was pregnant with her second daughter." The power of the story derives in part from how deeply transgressive is the image of a mother perched on a barstool for nineteen hours at a stretch: by implication, the woman is unemployed, unproductive, unfeminine, unmaternal, and morally degenerate.

Regardless of the source of their knowledge of FAS, doctors calibrated how much and what to learn about the diagnosis by how much they thought they needed to know. The pediatricians who regularly saw children with FAS had made the greatest efforts to educate themselves about the syndrome. Doctors who professed not to see patients with FAS were content not to enlarge their knowledge base beyond what they learned during medical school or residency. Many doctors suggested before or after the interview that I was "testing" them to see "how little we [doctors] know." "Oh, man, this is a test!" groaned one family practitioner when I asked him to describe the features of FAS. "What is the nature of your research, are you establishing that physicians don't know anything about FAS?" asked a pediatrician. Some doctors, in fact, asked me to tell them more about FAS.[16]

Doctors' Experience with FAS: "I've Never Seen It"

The doctors I interviewed may be divided into two basic groups: those who encounter FAS fairly regularly in their practices and those who reported rarely or never seeing patients with FAS. Doctors in the first group tend to be either pediatricians who specialize in treating "at risk" children or growth and development specialists. These doctors may see one or two patients a year with FAS, or they have under their care at any time a couple of patients with FAS. Some of them have long-term relationships with these children, treating them over the course of their childhood. One pediatrician said of a patient she had been seeing for about six years, "He's my N of one." Obstetricians in maternal-fetal medicine also tend to have more exposure to FAS (as well as other kinds of birth defects) than the average practitioner. They may recognize or suspect FAS in a newborn; however, unlike pediatricians, they do not have patients with FAS.

The second group consists of doctors who have never seen a case of FAS or who have seen a patient with FAS only once or twice in their careers. They may remember that case vividly, in detail. Often this exposure happened during their days as medical students or as residents. A few doctors said they had never seen FAS among their own patient population, but they had encountered it while working as a medical volunteer, often on American Indian reservations. Other doctors speculated that they might have "encountered a patient and not recognized it." "The reason I don't know more about it is [that] I have never actually [seen it]. Or if I have seen a baby with FAS, I don't think I would recognize it," admitted one family practitioner who had been in practice for some twenty years. About a third of my respondents reported they had never seen a case. Doctors who had only occasionally encountered patients with FAS and doctors who had never done so make up the majority of my respondents (17 of 30), suggesting that having experience with patients with FAS is by no means a routine occurrence for doctors. There are several explanations for this infrequency: doctors actually see FAS more often but do not know what they are seeing; patients are channeled to specialists early in life, thus a few doctors see a majority of patients; the condition is in fact rare; or FAS may be particularly rare in Philadelphia, although there is no particular reason to believe this is the case.[17] Moreover, it is important to recall that my sample is not representative of all physicians. If anything, the sample probably overrepresents clinicians' experience with FAS, because it includes a disproportionate number of doctors in maternal-fetal medicine and in developmental pediatrics.

Describing the Diagnosis: "A Small Baby That Looked Kinda Funny"

We expect medical diagnoses to be consistent across time, place, and instrument; what is a myocardial infarction, or heart attack, in the eyes of one doctor ought to be a heart attack in the eyes of his contemporary colleagues as well.[18] And when we ask doctors at a particular time and place to define a medical condition, we expect that their answers will be fairly invariant. We expect such consistency at least within cultures and historical eras. Medical knowledge, of course, varies dramatically over time; disease perception and definition can vary from one culture to the next among both doctors and laypeople.[19] Although doctors' individual understandings of the disease may be colored by their own clinical experience, we expect a core consensus. We think of medical knowledge as universalistic: all doctors — "specialized expert[s] in a specifically defined, if broad and complicated field"[20] — know the same basic things. Yet individual clinicians' understandings of FAS suggest that knowledge of this diagnosis is neither universal nor invariant.

Some doctors begin their descriptions by portraying the phenotype, cataloging the morphological changes in the child's face. "That whole physical facies kind of bespeaks it," said one pediatrician. It's "just something about the face," contended another. Other doctors focused initially on the fact of maternal drinking as the most salient feature of the diagnosis. Others described first the developmental delay, mental retardation, and behavioral deficits manifest in children with FAS. These varieties of descriptions showed no pattern with respect to specialty. Even within specialty, there are no universal features of doctors' descriptions of the syndrome. Certainly everyone is describing the same picture, but they are using different words to do so and focusing on different details and elements of that picture. As one obstetrician put it, "I think it is a confusing syndrome, but that is true for a lot of syndromes." Thus, while doctors' descriptions of FAS reveal many common elements, in many ways the diagnosis differs from one doctor to another. No single feature of fetal alcohol syndrome was mentioned consistently by every interviewee, although most doctors made some reference to maternal consumption of alcohol or in utero exposure to alcohol.

However, the epistemological relationship of a maternal history of drinking to the diagnosis of FAS was often a confused one. Some doctors suspected FAS based on a child's clinical presentation and then looked for a maternal history to confirm the diagnosis. "[The] types of things that would make me suspicious of FAS would be . . . abnormal facial features, if the child has developmental delay, some cognitive impairment, maybe behavioral problems, learning disabilities, frank retardation . . . you have to put the whole picture together and then you elicit that history," suggested an obstetrician [7B, female]. Other doctors used a known maternal history as a justification for seeking the symptoms of FAS in the child. "If I knew [the] mom had been drinking . . . I would know what to expect or look for. I would say, you know, I bet I am going to get a FAS kid," admitted one family practitioner. What the proper relationship of history to diagnosis should be is unclear. A maternal history of prenatal drinking is a requirement for the official diagnosis; the diagnosis of FAS should not be made without that history.[21] However, whether physicians should be able to "see" FAS in an infant or child without knowledge of maternal history and then use maternal history to confirm their suspicions, or whether they may properly use a maternal history of drinking to alert them to the otherwise subtle manifestations of FAS, is uncertain. One pediatrician mentioned a history of alcoholism in the family: "have to be equal, equal rights," he said. "I think there is some paternal influence." Except for this pediatrician, doctors in my sample believed FAS to be exclusively the result of the woman's behaviors during pregnancy.

Doctors described FAS as "not very pleasant," as "fetal exposure to toxins," as the "anatomic and intellectual consequences of excessive alcohol abuse during pregnancy," as a "complex syndrome of a variety of characteristics that express themselves in a complex variety of ways from mild to severe to full-blown syndrome to some FAEs [fetal alcohol effects] that have various manifestations that may or may never be picked up," and as "a clinical diagnosis of a certain pattern, thought to be brought on by drinking of alcohol during pregnancy, that results in certain dysmorphic features of children, that are also associated with behavioral and intellectual problems." Many doctors talked about babies or children with FAS as "funny-looking." Some doctors saw a "whole spectrum [of effects], although it is often very subtle, and, in fact, so subtle that without a history of drinking it may be hard" to recognize even in affected patients.

Funny was an adjective commonly used by doctors to describe FAS. "Just knew we had a small baby that looked kinda funny . . . It had sort of a weird facial structure," recalled one obstetrician of the first case of FAS he saw. Doctors referred to children with FAS as "funny-looking kids." "Funny face, funny hands, funny head," said one. "These kids can look pretty strange at times," said one doctor. "They look kinda weird," said another. FLK, or "funny-looking kid," is a catch-all diagnosis in medicine, used to describe subtle anomalies for which doctors cannot pinpoint a cause. That the word *funny* recurs so often in doctors' descriptions of FAS suggests that the formal diagnosis is a codification of what in the past was an informal diagnosis for an ambiguous, unclassifiable anomaly. In this light, FAS seems to be almost a variety of FLK, a formal subtype of an informal "diagnosis."

Certain emotionally significant words recur in doctors' descriptions of FAS. They speak of lives that are "dramatically devastated," of the "sadness" of FAS, the "tragedy" of the syndrome, the "damaged" children it causes. Such emotionally weighted words tend to be used more often by pediatricians — who are more likely to see and to sustain relationships with children with FAS — than by obstetricians. These words suggest the emotional resonance that the diagnosis holds for some doctors. Children with birth defects are characterized as "devastating surprises" for their parents. Many doctors see FAS as a particularly devastating diagnosis for three reasons in particular. First, doctors believe children with FAS are especially difficult to care for. "It is so difficult to deal with, they [the parents] are so overwhelmed by the problem, that it's all consuming." Second, a diagnosis of FAS carries with it a strong element of blame on the mother: "I think it would be devastating for a woman to learn that she has through her carelessness or indifference permanently damaged another, her own child." Third, the diagnosis implies the "unfitness" of the mother, her inability to mobilize the social resources necessary to cope with the stresses of her life or the responsibilities of motherhood. Doctors frequently mentioned the "dysfunctional families" of children with FAS. "They're very high risk families," said one family practitioner. Another talked about substance use in pregnancy as a "marker for catastrophic psychosocial issues." "Most likely the social situations that they [children with FAS] are born into are not the best nurturing environment," noted a pediatrician.

Pediatricians were most likely to see children with FAS as innocent victims; one characterized the vulnerable fetus "surrounded by toxins" as a "child who has no voice." Pediatricians reported feeling "great sadness: that this human being was, has been compromised largely because of either the mother's ignorance of the dangers or unwillingness to do something about the habit to protect her child." Pediatricians saw children with FAS as suffering because of the mother's "irresponsibility." "I think it is tragic as a pediatrician. I feel it is tragic for the child . . . that they are going to have lifelong consequences of what happened during their development" [12P, female].

Other doctors viewed FAS more dispassionately. "It's just one of many tragedies I see everyday," said an obstetrician who specializes in high-risk pregnancies [2B, female]. They saw FAS not so much as sad as "unfortunate," which removes some of the opprobrium from the woman who drinks during pregnancy as well as some of the sense of tragedy. In thinking of FAS as unfortunate rather than tragic, these doctors moved the syndrome along the spectrum of responsibility to a position closer to genetic disorders, which are seen as random, unpreventable occurrences.

Medical Uncertainty about Etiology: "I Don't Think Anybody Knows"

More confusing than the diverse clinical pictures of FAS described by the doctors is the range of etiological mechanisms that they postulated. No less than reading the medical literature on FAS, talking with doctors about the syndrome reveals considerable uncertainty about the etiological mechanism behind FAS. That uncertainty is visible in two ways. First, doctors' explanations of what causes FAS span a wide range. They attribute the syndrome to "excessive alcohol abuse during pregnancy," "drinking of alcohol during pregnancy," "a good amount of alcohol throughout pregnancy," "maternal alcohol consumption," "in utero exposure," "mother's intake of alcohol during pregnancy," "maternal ingestion of alcohol," and "substantial binge drinking or drinking daily during pregnancy." One doctor told me, "It is not entirely clear how much alcohol it takes to cause the different degrees of it [FAS], although it seems for the most part to have to be a substantial exposure." One family practitioner claimed, "I believe it can really be from even one glass to an abundant amount" [22F, female], but an obstetrician noted that "an occasional drink is not going to give the baby FAS." Despite this uncertainty, doctors commonly maintained that "this is very risky business . . . There is no known safe amount," thereby reiterating the U.S. surgeon general's official assessment of risk. Indeed, the range of answers given by doctors echoes the conflicting information in the medical literature about the risk levels for FAS.

These different formulations not only display a range of levels of alcohol exposure believed to cause FAS, they also assign gradations of responsibility to the woman who drinks during pregnancy. All these descriptions in some way implicate the woman, but they place greater or lesser weights on her role. Some doctors — especially pediatricians — invoke the woman explicitly through the words "maternal" or "mother's." In other formulations, more typically used by obstetricians, her actions are merely implied: "drinking of alcohol," for example, elides mention of agency.

A second indication of uncertainty is that, as some doctors readily conceded, we do not know what causes FAS: "We don't even really know what the method of action, the mechanism of teratogenicity, is. We haven't really figured that out." Doctors' comments echo the debates in the medical literature about not only the level of alcohol exposure but what exactly the causal link between alcohol and birth defects is. "We know more alcohol is likely to cause problems, particularly early in pregnancy. We know that some women who drink quite heavily manage to produce normal babies. And so there must be something else going on. Maybe it's — I don't know, who knows what?" said one pediatrician [10P, male]. "I don't think anybody knows [the etiology]. I think it's the way the alcohol interacts with developmental processes and I don't know that anybody knows yet the link" [20P, female]. One obstetrician replied that "fetal exposure to alcohol presumably" was the cause of FAS.

> It appears, since it is a spectrum, and since the changes are not always the same in every fetus, that it's probably a multifaceted method of action, if you will, of the alcohol. It's theorized that [it is a] direct or indirect effect of either alcohol or its byproducts, acetaldehyde. But no one's really shown a good cause-effect between either alcohol [or] acetaldehyde and each fetal abnormality. So it is just sort of presumed, and the method of action is thought to be multifaceted and not really pinpointed . . . I don't think there's an exact threshold that has been identified, there

isn't even really, as far as I can tell, sort of a dose-response curve that's been developed either. [21B, male]

This doctor's account of "what causes FAS" reveals considerable uncertainty: the cause is multifaceted, it is not pinpointed, it is theorized, it is presumed. There is nothing exact or precise about this explanation. The doctor's uncertain knowledge is reflected in the imprecision of his language. Moreover, this uncertainty is not his alone but is a facet of the collective medical understanding of FAS.

Since "the changes are not always the same in every fetus," doctors also speculated that some individuals may be more susceptible to FAS than others. "The trigger is in utero exposure to alcohol and probably susceptibility by some individuals or some pregnancies, and the amount of alcohol is varied. But [I] think [it is] alcohol as a teratogen or an alcohol-breakdown product as a teratogen" [26B, male]. Thus, doctors acknowledged several dimensions of uncertainty about the etiology of FAS: how much alcohol poses a risk; whether alcohol itself is the teratogen at work; who is at risk of experiencing exposure to this teratogen, whatever it may be; and what factors contribute to differential susceptibility to that teratogen.

Consequently, doctors arrayed themselves along a spectrum from permissive to prohibitionist in terms of the advice they gave their patients about drinking during pregnancy. In truth, however, doctors tended to cluster at one end of the spectrum or the other. Most obstetricians generally adopted permissive attitudes towards drinking during pregnancy. Whether it was because of their experience advising use of alcohol therapeutically in pregnancy or because of their concern not to alarm women unnecessarily, most obstetricians reported counseling women that "once they get out of their first trimester, if they want to have an occasional drink, that's OK" [17B, female]. "Now I tell patients, for example, New Year's Eve, if they happen to have a glass of champagne or a glass of wine, don't panic out . . . , but in general try to avoid" alcohol [15B, male]. Thus, obstetricians tried to be sensitive to their patients' lives outside their pregnancy, recognizing that pregnancy does not preclude other activities in women's lives.

Pediatricians, in contrast, tended to be more absolute in their beliefs that women should abstain from any drinking during pregnancy. "Of course, it is probably true that a small alcohol intake is not going to harm the baby, but we know that alcohol has absolutely no beneficial effects, we don't know what a safe amount is, so the wise course is don't touch it, don't do it" [10P, male]. Doctors like this pediatrician responded to uncertainty about risk with absolute prohibitions: "Don't touch it, don't do it." Like pediatricians, obstetricians who specialized in maternal-fetal medicine tended to take a more cautious approach towards drinking during pregnancy. "I tell them [patients] that abstinence is the best policy because we are not sure of a threshold. In other words, clearly, taking more than, you know, three to five ounces or so of alcohol a day is bad, so is two ounces bad? Is one ounce bad? Is a half-ounce bad? There is no threshold, so that the only thing I can tell a patient that is completely safe is no alcohol use" [14B, male]. Another maternal-fetal specialist said, "I think any level is probably risky. I think the question I would turn around the other way and ask is there any level I could identify as safe?" [26B, male]. Thus, the extent to which physicians saw themselves as responsible for the well-being of the fetus or child colored their perceptions of risk: the more they focused on the fetus or the child, as opposed to the woman, the more cautious was their assessment of risk.

Family practitioners, on the other hand, expressed a range of attitudes towards drinking, from strict to permissive. "Well, I don't really know, I don't really know [how much drinking is risky]. But I can't believe that [an occasional drink] does a significant amount of harm, even though there are some papers that say it may," said one family practitioner [29F, female]. Another counseled patients by saying, "I don't know how much is the total quantity you can have. I can tell you that what I know [is] you are not supposed to drink during pregnancy, but if it's Easter and you are going to have a glass of wine and you have a sip, well, more power to you" [30F, male]. "I tell them abstinence is the best policy," another family practitioner [18F, male] reported.

Making the Diagnosis: "It's Pretty Subtle, Yeah"

Just as doctors did not agree on what causes FAS, they differed on how many children are affected and on how easy it is to recognize those children. Spotting the diagnosis can be confounded by the psychosocial features of the families into which children with fetal alcohol syndrome are born. Doctors perceived the effects as being "very subtle, just behavior findings that are blown off to something else. You've already probably got a dysfunctional family and a lot of it is possibly blamed on that" [26B, male]. The delayed development and behavioral problems that children with FAS exhibit are easily discounted by doctors as products of their environment. "I think it is complicated by a lot of different things, but probably the alcoholism is underdiagnosed in the parents and then in moms and they're often in dysfunctional settings, so you know, for there to be behavior problems in those kids, who are from single parent families who are on welfare . . . is normal . . . it doesn't end up making red flags go up about referring those kids for further evaluation" [18F, male]. This physician's remarks reveal a lot about who he thinks is susceptible to FAS: in his mind, the syndrome is a problem of poor women, single mothers, women on welfare. Many doctors, in fact, perceived FAS as enmeshed in a broader web of social dysfunction shadowing a family's life.

Other doctors thought they or their colleagues might hesitate to make the diagnosis, because "it's a big deal to make the diagnosis of FAS," in the words of an obstetrician–clinical geneticist [2B, female]. "There are major implications for a family if a diagnosis is made . . . It's labeling a woman as having caused significant morbidities to her child through her own illness, which in many sectors of society is viewed as her own indulgences," noted an obstetrician [14B, male]. One doctor suggested that the subtlety of the syndrome's markers combined with the potential impact on the family made him "want to be very sure" about the diagnosis. "It creates some huge issues of blame, guilt, and dynamics between an affected child and the mom with reference to a lifestyle choice that not uncommonly is already embedded in the family system and there are already issues with an alcoholic mom." This family physician saw the families of babies with FAS as already particularly vulnerable; making a diagnosis of FAS could further disturb a precarious situation.

Another family practitioner reported a complicated calculus in weighing whether to make the diagnosis or not. His decision was influenced not only by the physical manifestation of the syndrome in the child but by his sense of the effect of the diagnosis on the family, as well as what he called "the nebulousness of making that definitive diagnosis." "You're weighing . . . personal issues and costs and guilt and blame issues on the one hand, against not being able to have a great gold standard for diagnosis on the other hand. So if I am going to get into issues with the family, I'm going to be wanting to be darn sure that I know what I have. And usually I want to have something that I can do for their sake, and that's kind of a gray zone here" [23F, male]. Thus, the diagnosis of FAS offered few advantages to this doctor and in fact had the potential to cause more harm than good. He saw the diagnosis as invoking "guilt and blame" and worried about its effect on family dynamics. He also found the lack of a "gold standard"[22] for diagnosis unsettling: before making a diagnosis with such high potential costs, he wanted to be sure he was correct. But with fetal alcohol syndrome there is no way to be sure. Finally, he found the diagnosis lacking in usefulness because making it does not increase the treatment or ameliorative options he could offer to the family: there is nothing he could "do for their sake."

Other interviewees echoed these sentiments. "I think one of the other reasons that it probably isn't more aggressively diagnosed is that there is no effective treatment in terms of reversing the damage." Thus, doctors saw a certain futility in making a diagnosis of FAS in the absence of anything they could do medically for the children so diagnosed.

Still another reason offered for the presumed underdiagnosis of FAS was some form of ascertainment bias. "We are not used to looking for it here [in Philadelphia]," conceded one doctor [13F, male], contrasting the attitude at his hospital with that he had observed on American Indian reservations. Another recognized that doctors themselves are not always the most sensitive of diagnostic instruments. "It depends on what population is doing the diagnosing, I suppose. I think in general, general pediatricians underdiagnose causes of things and deal more with the things themselves and probably take a more hands-off approach than, say, you would in an academic center. We see this in the clinic all the time, when kids come here who are very small and the pediatrician will say, well, that's just the way he was meant to be" [9P, female]. However, the ascertainment bias runs deeper than mere insensitivity to the syndrome and its symptoms. As one doctor noted, "There are certain populations where physicians would not consider the diagnosis of FAS." Doctors see and find FAS where they look for it, and they may be inclined to look for it only in populations they stigmatize as taking risks during pregnancy. An obstetrician said, "I bet probably in high-risk groups, such as Native Americans, they probably overdiagnose it . . . Because, I mean, these people are reported to have a high incidence of alcoholism . . . FAS may get more play in areas where it's more commonly found" [28B, female]. Likewise, "the history and the current behavior of the child" may influence whether the diagnosis is made or not.

> Like, if you have a child that's not doing well in early school and they are an ADD [attention deficit disorder] child, for example, and they look a little strange, I think somebody would more rapidly say FAS, whereas if they were doing well in school and they just had a little strange [shape] of a jaw, they [the doctors] would [say],

oh, it's just a strange jaw . . . So I think in practice it is harder to see and I think that it is a biased diagnosis . . . If you're a successful kid and you just have some small characteristics of FAS, no one is going to say you have FAS . . . but if you have a kid that has fallen back in reading class and is behaviorally difficult, and he has wide-set eyes, then some pediatrician, you know, is going to say "classic FAS" . . . The mom drank, or the kid's a difficult child, and so then all of a sudden, those physical characteristics become more evident. [30F, male]

The symptoms that the doctor "sees" in the child are conditioned by what he sees around the child, in her environment and history.

Finally, doctors believed that a large number of children affected by prenatal alcohol exposure slipped through their diagnostic nets. "Because I think — you know, we all look for the classic symptoms or visible features. And most syndromes have a spectrum and my suspicion is, for example, as I said, I think more babies are probably more adversely affected from a growth point of view by alcohol than the full-blown syndrome, and so we are probably missing some of the spectrum, the in-betweens."

These beliefs about the propensity of their colleagues to underdiagnose FAS, combined with their overestimates of the incidence of FAS, suggest that many doctors believe FAS and FAE are substantial sociomedical problems. That assessment is all the more curious, given the lack of experience with FAS among many doctors in the sample. It is almost as if the less FAS they see, the more they think it exists. In fact, doctors today may be somewhat conditioned to overestimate problems with social roots — like child abuse or domestic violence — as a correction to the almost total invisibility that historically has cloaked such problems in clinical medicine.[23]

Curiously, doctors did not agree on whether the diagnosis is a hard or an easy one to make. Given the lack of a biological marker, the purported subtlety of the features, and doctors' avowed belief that the syndrome is underdiagnosed, one might expect doctors to believe that the diagnosis is not easily recognized. Many doctors certainly believed that to be the case. "There's nothing really absolutely objective that we can use to say this is FAS. It's not like a blood test that you can do and look for a particular enzyme or protein or chromosome changes. There's nothing that you can really hang your hat on except that you have to put the whole picture together and then you elicit that history" [7B, female]. This obstetrician's description of the diagnostic process may reveal a common premise of doctors' diagnoses of FAS: you identify first what you believe to be the sequelae of maternal drinking in the child, then you seek out confirmation that prenatal alcohol exposure occurred. This reasoning process, whatever its clinical rationality, nonetheless introduces a serious logical flaw into our understanding of the relationship between alcohol and pregnancy, since it looks for a history of prenatal alcohol exposure only in children believed to have FAS. Clearly, some doctors were discomfited by the lack of a definitive diagnostic test for FAS. "Part of the problem is [that] without a biological marker, it gets very murky." Even aside from a biological marker, there is "not a specific anomaly, unfortunately" that allows doctors to stake a sure claim to diagnosis. "You know, no kids say 'F.A.S.' across their foreheads," said a family practitioner who reported that he had never seen the syndrome among his own patients.

The word *subtle* recurred throughout the interviews; doctors believed the diagnosis is "pretty subtle." "I think it's sort of the nature of the syndrome [that] makes it hard to diagnose. Some of the changes can be subtle, and someone who is not trained on what these kids look like could miss it or could call it something else" [21B, male]. "If the face is all swollen from a difficult delivery, you may not see the features. Some of them are fairly subtle . . . I am not even sure I know what they are," admitted one pediatrician [10P, male]. In fact, the elusiveness of the syndrome was another reason doctors advanced for its underdiagnosis.

Yet, other doctors saw FAS as a clear, unmistakable diagnosis: "Its description is fairly precise as far as the characteristics the child has, and clearly when you see children that have FAS, it's right there" [15B, male]. These doctors claimed that the syndrome is "characteristic and classic," that it presents "a clear profile." One doctor with experience working on an American Indian reservation talked about seeing clinic waiting rooms filled with children with FAS — "they kind of all look the same . . . you could see them running around all the time. It is so common, you realize that something is going on, either genetically or something, and it wasn't long before you realize this was fetal alcohol [syndrome]" [13F, male]. Another doctor recalled a trip to Kiev — "just walking down the street you get the feeling you see a few" adults with FAS. "Just something about the face" indicates the diagnosis to him. His impressionistic sense of fetal alcohol syndrome in strangers passing by on the street illustrates the subjectivity with which doctors view FAS. Like the range of descriptions they offer and the range of causes they hypothesize, doctors' varying assessments of whether it is hard or easy to make the diagnosis epitomize the slipperiness of this diagnostic category.

Using the Diagnosis: "You Have a Label but Not a Better Outcome"

Some doctors hesitate to make the diagnosis of FAS because they believe it is not clinically helpful. How, then, is the diagnosis of FAS useful to physicians? Does it provide a clinically and intellectually useful category? While few doctors in the sample challenged the diagnosis itself — most seemed to think that FAS truly exists in some sense, that the diagnosis corresponds to some real entity — many nonetheless questioned whether making the diagnosis offers any benefit to doctor or patient. "I am not aware [that] having made the diagnosis, the treatment options are dramatic to the point that you can make a difference. So it is more like labeling a chronic condition and then what do you have? You have a label but not a better outcome" [23F, male]. "Well, it identifies what the source of the problem is. It does put together a group of findings into a meaningful pattern. Can't do anything about it. Doesn't help from that point of view" [10P, male]. "What does making that diagnosis change?" asked a family practitioner. "Besides putting a stigma on the child, it also stigmatizes the mother." Because of the ambiguity of the syndrome, "what real gain is there in forcing a diagnosis if there isn't one?" Rather, some doctors worried about the potential of FAS to label children, to stigmatize them and their mothers. "I don't like to see kids labeled, that is my objection," said one obstetrician [7B, female]. "Labeling for its own sake is just an academic exercise," contended a pediatrician [19P, male].

Moreover, the diagnosis of FAS carries real peril for the mothers of babies and children so labeled. "It identifies her as a person who's done a bad thing to her baby, you know. There is a stigma that goes along with that." The diagnosis of FAS also threatens to disrupt already fragile family dynamics.

> I actually try to discourage these families from continuing to bring that [a FAS diagnosis] up, because I do — I do think it is something that is done, is past. I think we cannot affect nature, now we have to affect nurture and I think it is time wasted talking about that diagnosis for these kids, because whatever is, is, and I think that, you know, now we have to deal with early interventions and all of the things that foster families can give these children that will then overcome whatever has hap-

pened to them in nature. So I don't think that having the full diagnosis has helped anyone, except for perhaps — actually, I'm not sure it has helped anyone. I guess it has given some explanation, maybe, for some small stature and microcephaly, but we do not further have to beat that diagnosis into a pulp . . . because there is nothing you can do about it. [12P, female]

Again, the futility of the diagnosis seems to defeat any benefit there is to making it. For the most part, obstetricians seemed primarily concerned with the stigma that might attach to the mother when her child is diagnosed with FAS, while pediatricians worried more about the negative effects of labeling on the children. Many doctors simply regarded a formal diagnosis of FAS as more harmful than helpful.

Many doctors, however, did identify a variety of ways a diagnosis of FAS could help children, families, and clinicians. This aspect of the diagnosis is an accepted item of faith in the medical literature. "Appropriately diagnosing alcohol-related birth defects in these children can be a great relief to parents and teachers and can serve as a basis for implementing remedial programs to minimize the negative consequences of the developmental deficits."[24] In its potential to explain a child's difficult behavior or to link a child to needed social services or developmental interventions, the diagnosis is a force for good in the eyes of some doctors. Many of these reasons are related to outsiders' perceptions of the child. Namely, doctors believed that the diagnosis can change others' perceptions of the child and his behavior — "I think it would help folks be more empathetic" — and it could help the family get the services they need. "I think it can help them get other services. You know, people take it more seriously, as compared to your saying, 'I think this kid is at risk.'" Doctors also suggested that the diagnosis of FAS could allow social workers and teachers to tailor their interventions, to prevent the mother from having another pregnancy affected by alcohol, and to link the family to appropriate support groups. However, other doctors noted that developmentally delayed children gain access to such services based on their problems, not on specific disease labels. A family practitioner said, "there is no treatment that a FAS child would get that a developmentally delayed [child would not get] . . . The same services that would be available to a non-FAS developmentally delayed child are available to an FAS child, so really the interventions are based on their development, they are not based on whether they have FAS or not."

However, perhaps the most important justification doctors offered for the diagnosis of FAS was its potential to provide an explanation to a family suffering the effects of the syndrome, to help them understand what is going on. "The family and the child would have a handle, you know, sort of a reason, an understanding of what was happening to them, why things were going the way they were going." Rather than exacerbating family tensions, the diagnosis can alleviate them. "Well, I think recognizing that a child has an organically based disability as the root cause for the child's intractable behavior is of some help." In one pediatrician's experience, the diagnosis "helped parents [especially adoptive or foster parents] in giving credence to the idea that it wasn't just how they were raising him that was a problem . . . When kids have pretty severe behavior problems, the parents are always the first suspects as the cause for that" [9P, female]. Since many children with FAS are adopted, some doctors saw it as especially important to absolve adoptive parents of responsibility for having "caused" their children's maladaptive behavior.

The FAS label thus has the power both to apportion blame and to erase it. Finally, it "allows you to be more predictive of what the future may hold . . . It allows you to counsel the family about what to expect for their child" [11P, male]. Nevertheless, the explanation afforded by the diagnosis could also introduce more suffering into the family, particularly considering the potential of the diagnosis to induce guilt in the mother and the very limited prognosis for the child. One pediatrician reported that she felt "a sense of dread" when she had to make the diagnosis of FAS, "knowing that the complications for this child are going to lead to further frustrations on the mother's part, that it's going to feed in and it's going to affect the child negatively" [8P, female].

In fact, doctors emphasized the "terrible" prognosis for children born with fetal alcohol syndrome. These families could expect little but hardship for their children. Formal diagnosis or not, doctors believed these children would "lead a hard life in the long run." Given the dire circumstances of these births, many clinicians felt powerless to intervene and change outcomes in any major way. "So here we have an alcoholic mother with a problematic child. It's just a risk factor for major problems," noted a family practitioner [1F, female]. The problems that "caused" FAS, as well as the problems that the syndrome itself "causes," seem to overwhelm the clinical need for a diagnosis. Doctors often feel helpless to address either the symptoms or the situation.

Doctors' Feelings about Mothers and Children:
"I Suppose a Combination of Sad and Angry"

"An alcoholic mother with a problematic child": this summarizes the way many doctors viewed children with FAS and their mothers as clinical *and* social problems. Both the child with fetal alcohol syndrome and the woman who is an alcoholic pose particular clinical challenges to physicians, since there is little in the way of therapy that medical doctors can offer for either patient. How do physicians view such patients? What feelings do such women and children elicit? Doctors admitted to feeling anger, sadness, and frustration with these patients. They reported feeling "frustrated for [women who drank during pregnancy], frustrated *at* them and frustrated *for* them" [16F, male]. Some doctors overtly blamed women for FAS in their children; other doctors avowed that it was not their place to judge women. Many doctors sought to place these women and children in a social context that provided an explanation for the tragedy they saw. They believed that women who drink during pregnancy "couldn't help themselves." On the whole, obstetricians and family practitioners seemed least judgmental towards women who drink, and pediatricians seemed angriest at such women for the "damage" they caused their children, although even many pediatricians expressed compassion for women who drink. This pattern of concern fits the differences in types of patients each specialty considers itself responsible for.

Almost universally, doctors expressed sadness for the children. One pediatrician said he felt "great sadness, that this human being was, has been compromised largely because of the mother's either ignorance of the dangers or unwillingness to do something about the habit to protect her child" [10P, male]. His sadness, however, is not unmixed with disapproval of the mother. Other doctors expressed sadness for both child and mother. "[I] certainly felt bad for the kids because they were really suffering from something they didn't have any control over . . . I also feel horrendous for the parents . . . That's probably one of the more difficult, frustrating diseases to take care of a kid with," said an obstetrician [26B, male]. Some doctors saw the whole family as suffering. "It was sad. To see these poor kids destined to be troublemakers, get the short end of the stick because their mother, who also got the short end of the stick, who probably got the short end of the stick from her mother — you know, these are generational problems, years of dysfunction. So it was sad"[25] [16F, male]. Thus, this family practitioner's feelings were both tempered and intensified by his sense that FAS is related to social inequality — "getting the short end of the stick." His comment is tinged with fatalism, with a sense that there is little he can do to ameliorate "years of dysfunction," or what another pediatrician called "the overwhelmingness of their [the women's and children's] lives." It is both the women and the children who suffer the effects of FAS *and* the doctors who are called upon to treat the syndrome and its causes who are overwhelmed. FAS thus serves as a paradigmatic example of how social problems can defeat physicians' efforts to improve their patients' lives and vex their feelings of responsibility towards their patients.

Doctors hesitated to express openly their disapproval of women who drink. Some seemed genuinely compassionate towards women who drink and the troubled lives they lead. Others chose their words carefully and deliberately. Rarely did doctors express open opprobrium for women who drink while pregnant. One obstetrician acknowledged that "your first reaction is to lash out at the mother, blame her." However, he went on to say that after years of working with women, he had come to appreciate that most women take the responsibility of pregnancy quite seriously. "I think it is just incredible the commitment women have to babies, and that can be anyone. It is not restricted to the educated middle class or upper class. It is across the board. It is a rare patient that really doesn't care about that pregnancy" [15B, male]. However, some doctors did find fault with the woman's behavior, particularly drawing a line between genetic malformations, which they saw as random events, and the effects of fetal alcohol exposure, which they saw as the consequence of the mother's actions. One doctor noted, "with most genetic diseases, you spend a lot of time discussing the fact that the patient's not to blame for it, for a genetic problem, these are random [events]. And you know in this situation you are talking about, yeah, somebody is to blame and it becomes very difficult" [2B, female].

Some doctors confessed to feeling anger towards the women who give birth to babies with FAS "because of what they have done by their lack of care or ignorance. [It] doesn't just impact on them, it impacts on their babies." Embedded in their denunciation of these women is the notion of the fetus as an innocent victim. "You feel sorry for 'em [babies with FAS], largely because you realize they have an illness that's going to impact on them for the rest of their lives, that they were [not] predestined to have, that was not of their own doing, that could have been prevented" [17B, female]. One family practitioner said he had seen pediatricians express open contempt for children with FAS. "These pig-headed pediatrician who think they're gods — [they] just spit on kids with FAS." Yet few doctors admitted openly in interviews to such feelings. Occasionally, I could hear unexpressed feelings manifest themselves in the interviews, as when a family practitioner found himself wondering with anguish about women who have babies with FAS, "How did it happen? Why didn't you care?" A family practitioner described how he thought others felt: "To other people . . . it's worse than alcoholism, 'cause it's alcoholism when you're pregnant. It's like exponential alcoholism: not only are you a drunk, but you are a drunk when you are pregnant. Dreadful, you know. Horrible member of our society, and it is very judgmental. That's what I think the stigma is. It's, like, worse than a drunk" [30F, male].[26] Doctors seemed willing to forgive a woman the first time she has a child with FAS, but they were especially angered by women who have multiple births affected by alcohol. "When the parents come back again and again, I don't have nearly as much empathy with them and I feel angry when I see that occur, when people actually know that they are bringing in kids who are damaged. That pisses me off" [26B, male].

Other physicians, however, seemed genuinely compassionate and concerned about the disordered lives that produce such tragic outcomes. "Well, it is hard to criticize the people for what they do unless you can live their lives" [13F, male]. Many doctors reported that they thought it was not their role to pass judgment on women or their choices. "I have dealt with a lot of parents, a lot of women who are in very horrible social situations while they're pregnant, and I guess I reserve judgment on those people. I think that it is — life is hard, and harder for some than others. And I think that if you could sit down and speak with those people, I bet there was a lot going on" [12P, female]. In recognizing that "there is a lot going on" in the lives of these women, this pediatrician implicitly acknowledges that FAS in their children is not the result of the mothers' "indifference" or "carelessness." Rather than a license to judge, doctors saw a mandate to do what they could to help these families. "I don't think it is my role to lecture them or to tell them it is wrong or right, but rather to try to intervene . . . rather than being scornful, I tend to find that fairly pathetic behavior," said an obstetrician [14B, male]. "I have learned, have come not to blame or feel anger at parents in most situations for what's happened to their children, because I think in most situations parents are not necessarily fully understanding of the consequences of what they do . . . I've learned not to place blame. It is not my job to find them guilty. My job is to work with them for their child" [11P, male]. This pediatrician puts his duty to help the child and the child's parents ahead of an agenda of assigning responsibility for adverse outcomes. In fact, doctors frequently expressed a kind of resignation in dealing with the problems that their patients presented.

In striving to feel "more of an empathy or more of a sense of 'what a shame'" rather than anger, doctors sought to locate FAS in a social context, to see it as more than the result of individual behavior alone. One pediatrician acknowledged,

> I guess it's sort of a sense of anger, that she has, that a problem of hers affects someone else so significantly, and a sense of dread knowing that the complications for this child are going to lead to further frustration on the mother's part, that it's going to feed in and it's going to affect the child negatively, and I think I probably feel more strongly about it now because I am pregnant and so I have a sense of, you know, before it was only a sense that I wouldn't do [that], but now it's sort of a very strong feeling that how could you expose your child to an addiction of yours? I mean, on the other hand, a lot of the mothers are also abused and in poverty so there is not just pure anger, it's sort of a sense of hopelessness and frustration with their situation as well. [8P, female]

Note that this doctor does not say she is angry at societal inequality. To the extent that she is angry, she is angry at the women who drink, even as she struggles to understand the context of their lives and behavior.

Two Groups of Patients: "That's the Spectrum You Are Dealing With"

Doctors varied widely in their opinions about who was at risk of drinking during pregnancy. Many doctors believed "it runs through all socioeconomic classes," but others did not hesitate to locate drinking and other risk-taking behaviors predominantly in the lower socioeconomic strata. However, regardless of how doctors viewed the relationship between socioeconomic status and risk, they tended to divide patients into a number of dichotomous categories: private versus clinic, informed versus ignorant, abstainers versus drinkers. The crucial distinction in doctors' minds is between the women who do not drink and do care obsessively about pregnancy and the women who do drink and do not (seem to) care about managing risk in pregnancy. Some doctors saw these groups as falling into neat socioeconomic categories; others believed women in either group could belong to any class. Thus, doctors divided patients into two basic groups: those at risk of FAS and other adverse outcomes due to their risk-taking behavior and those not at risk, because of their concerted efforts at risk avoidance. In doctors' eyes, patients range from women who are highly motivated and risk averse to women "who are not knowledgeable or really don't have the ability to change what they do." However, it might be more accurate to say that patients cluster at the extremes of this range. "My perception is that women here tend to belong to two classes. Women who are very motivated about pregnancy and don't drink, or at least won't admit to drinking, and women who kind of do all the risk behaviors and who are, you know, kind of lucky to be alive" [2B, female].

> I think the irony is that the people who most need to be concerned are maybe the ones who are least likely to be attentive to it. I mean there are people who — you know they are absolutely scrupulous and touch nothing and that's at one extreme, and then at the other extreme are people for whom being pregnant makes no difference at all in terms of their behavior. [19P, male]

> You've got one population that, they see something in the paper, [something] that comes from any legitimate newspaper, they are immediately going to go to the extreme of total avoidance or whatever, or they panic out because at some point during the conception phase they had one drink and they are sure that is going to be a major problem. To the other extreme, where there are patients that are consuming alcohol [but] aren't telling you they are consuming alcohol and you are not getting through to them, so that is the spectrum you are dealing with. [15B, male]

> I think there is not enough concern for FAS in the general public. But I think there's too much concern for FAS among people who don't need to be concerned about it. Which goes back to the people who aren't drinking that are worried about sips. They're too concerned, and the converse to that is the people who are still drinking aren't enough concerned. [30F, male]

How doctors placed patients in these different groups varied widely, however. Some doctors characterized patients by socioeconomic status and correlated low socioeconomic status with high-risk behavior. Others recognized that risk-taking behaviors could span the social scale: "it crosses all socioeconomic strata, there's not one, I think, that's at greater risk than another. They're all at risk." Not uncommonly, doctors acknowledged that the stereotypical profile of the woman who drinks during pregnancy is a lower-class woman; however, a few doctors thought risk-taking is actually more prevalent among more affluent women, or what one family practitioner referred to as "the cocktail circuit at the top of the pile." As one pediatrician put it, "There are a lot of rich people who are drunk." So while some doctors believed that drinking during pregnancy is "a problem that you see disproportionately in certain populations," others felt that "it can happen to anybody."

Regardless of how they saw their patients, doctors struggled to find explanations for their behaviors. How do doctors perceive the behaviors they observe in their patients? What explanations do they craft for their patients' actions? Doctors may "explain" drinking and other risk-taking behaviors during pregnancy in a number of ways: as a consequence of ignorance, as carelessness, as irresponsibility. Conversely, they see risk avoidance during pregnancy variously as an obsession, as hysteria, as an expression of love. The explanation that doctors choose colors their view of patients, their interactions with patients, and the ways in which they view their own roles and responsibilities as physicians. If doctors believe their prenatal patients drink because they do not know the dangers of doing so, their duty is clear and straightforward: to remove patients' ignorance by informing them of the risks. If doctors believe patients drink because they are mired in "such dire circumstances that they can't be concerned beyond sort of the next immediate thing they are going to do," they see their responsibility more broadly. In fact, they see a social mandate: "We as a society should do more about providing help to people that are likely to have that kind of trouble, because I think they're likely to have a lot of kinds of trouble and the other kinds of trouble are likely to have adverse outcomes on the kids" [9P, female].

Pregnancy and Control: "The Most Accomplished Pregnant Person That Ever Lived"

Many doctors see a certain group of patients as being overly concerned about risk during pregnancy — "they just kinda go nuts about their lifestyle, about trying to kind of maximize everything, like they get the right exercise, they get the right sleep, they get the right amount of sunshine per day." Some women, doctors complained, "don't want to take a Tylenol" while pregnant. "They are worried about the smallest amount [of alcohol], and I think, jeez, have a sip!" These women, whom they characterized as well-educated, professional women of high socioeconomic status, worry excessively about controlling all kinds of risk, not just drinking, during pregnancy.

> I think . . . some of the information has been taken on by the middle-class patients to such an extreme. You know, don't take a Tylenol, don't have a glass of wine at all when you are pregnant. Don't fly on an airplane. Don't do this, don't do that, as if you are just going to sit in a room for twelve weeks and then. I think it is our culture, because I know that people in Europe are a little more relaxed about some of these issues . . . People are too obsessed with not doing many things in pregnancy, including alcohol. I mean, they are obsessed with, you know, if they walk by a lawn that has — that their neighbor has lawn chemicals. I think that certain segments of the population are terrified that environmental things are going to affect their baby. Even if it is eating, you know, a certain piece of meat that may have a hormone, and that's a small group of patients, I would say [28B, female].

Such women are the pregnant equivalents of what Renée Fox has called, in another context, "the worried well."[27] They experience a kind of "hypersusceptibility to risk."[28]

This hypersusceptibility is motivated at least in part by some women's ambition to have a perfect child. I do not use the word *ambition* lightly: doctors often characterized these women as being driven, extremely motivated about their pregnancies. "These are determined people. [They say] I want to get pregnant when I want to, and I am going to do everything to get the best possible child out of it," noted a pediatrician [10P, male]. An obstetrician commented, "I think it's symbolic of the generation of women having babies today . . . I think that because . . . people have fewer and fewer children now, that everyone wants a perfect baby. I mean, there is this perfect baby, and people will settle for no less than the perfect baby" [28B, female]. "They want to do everything perfect to make sure that they are going to have the perfect child." This attitude objectifies the "product" of conception, both raising the expectation of an unflawed outcome and deepening the disappointment when the carefully nurtured and controlled product fails to turn out as desired.

Often, such women have deferred childbearing for many years or have had trouble conceiving. They have exercised control over the timing of their pregnancies and expect to exercise the same degree of control over the outcomes of the pregnancies as well. For such women, drinking—along with a range of other behaviors—becomes the locus of excessive concern, an area in which women attempt to exercise control over the outcome of pregnancy. As one family practitioner put it, "They are going to be the most accomplished pregnant person that ever lived." In fact, this pattern of behavior during pregnancy replicates a particular set of middle-class attitudes towards health, in which control, self-abnegation, and puritanism are dominant.[29]

To doctors, this self-control can seem excessive and misplaced. These women's beliefs fall into the category of "popular health superstitions."[30] Women worry about things that are not true risks. Or they carry their concern to such extremes that it becomes dangerous to themselves or their babies or both. One doctor recalled that her asthmatic patients sometimes have refused to use their inhalers: "Not being able to breathe has got to be worse for your fetus than your inhaler!" she cried in exasperation [27B, female]. Another doctor said that nutrition and weight gain are sometimes problems in upper-class women who are overly concerned about their appearance and obsessed with body image. One doctor spoke with annoyance of patients who asked him, "'Can I continue to walk?' Go ahead, walk, it's good for you! 'Don't I need bedrest?' Not right now, go to work! Do what you want! 'Can I drive to New York? It's a long drive.' Drive to New York, I don't care! You're not a social pariah because you're pregnant. 'My grandmother has a cold; can I go visit her?' Go visit your grandmother! Just wash your hands, for God's sake! Stop asking me these questions. So they're, like, hypervigilant. How do you think your great-grandmother survived without me? Just fine! You're here!" [30F, male]. Some patients have asked if they could take the communion wine at church on Sunday. Another doctor noted that women irrationally fear that anything could have "an everlasting effect on [the] fetus. Fortunately, most things don't, because if they did, none of us would be here! It's completely ridiculous . . . it doesn't make any sense" [27B, female].

Doctors noted that such fears are out of proportion to the actual risk. "I think the people that don't touch anything during pregnancy are probably overshooting." Doctors felt that some patients exaggerate the risks of everyday activities. "People are really afraid to do anything—to have half a glass of wine, to take Tylenol, to drink a cup of coffee. It's things that are done *excessively* that are a problem," said an obstetrician [27B, female]. This exaggeration of risk of everyday activities carries over into patients' attitudes about alcohol as well. One obstetrician recounted with horror the case of a patient whose fears had "gone overboard." The woman had undergone infertility treatment for eight years, "spending half her life's fortune" in a fruitless effort to conceive. After she and her husband gave up trying, they took a long cruise and when they returned, the woman found that she was pregnant. "She came in [to see me] 'cause her doctor recommended that she terminate the pregnancy because she drank on the cruise! Despite the fact that she had been [through] eight years of infertility [treatment]. And she had nothing, she had a couple of drinks a day! . . . She was very misinformed and some people overexaggerate the risks. She had a nine pound, eight ounce, kid, so I don't think it was a problem" [26B, male].

These kinds of excessive fears, although easy to laugh at, are but extensions of the sense of weighty responsibility that all women are made to feel in pregnancy and in maternity more generally. Pregnancy is regarded as a sacred state in the United States.[31] A few doctors noted the differences between American and European attitudes towards drinking in pregnancy, noting that Americans are "much more sort of hyped up about alcohol during pregnancy" than Europeans. "I've had patients from overseas who have commented that it seems like Americans are much more strict about alcohol during pregnancy than non-Americans, or at least Europeans," noted one doctor. Obstetricians in particular were attuned to the moral valence of pregnancy in our society. "I think the problem so often is that women are guilt-ridden by whatever they do during the pregnancy," acknowledged one. Guilt was another theme frequently invoked by doctors when discussing FAS and other risks during pregnancy. The disproportionate guilt may be the dark obverse of our society's obsession with control: since women are expected by others as well as themselves to manage risk in pregnancy, they are to blame when things go awry. Today, few outcomes are seen as accidents of nature.

A few obstetricians speculated that women's exaggerated fears spring from the kinds of popular health information current today. "I think that there is a lot of information in the media. They read a lot about pregnancy issues in women's magazines. There is a lot of advertisement everywhere about drinking and being pregnant. And I think that there is this general notion about, you know, trying to do everything to protect the fetus and how you shouldn't do anything when you are pregnant. I mean, some patients will not even take a Tylenol. It doesn't come directly from, I don't think it comes directly from the health care provider, you know, [the idea] that everything is bad" [28B, female]. Two family practitioners specifically mentioned the popular pregnancy manual *What to Expect When You're Expecting* as contributing to patients' worries about pregnancy.

Doctors perceived that many women view pregnancy as a significant trust. "But women are so guilt-ridden about being this vessel or container for a pregnancy that most women — and I think there are few exceptions — really take that responsibility quite seriously." Their sense of responsibility leaves them vulnerable to punishing feelings of guilt in the event of a bad outcome. Doctors were aware that women could be subject to extreme guilt if something goes wrong with their pregnancy. "You don't want to be guilty. If something went wrong, you say, 'Must've been that one glass of wine last Valentine's day'" [6P, male]. "Mothers have felt guilty for a lot less than this" [7B, female]. "I think most women feel a tremendous sense of responsibility for what they do or don't do in the pregnancy . . . Any bad outcome is what I should or should not have done. I did something wrong. They immediately jump to that conclusion and everybody around them will support that guilt trip, so I think women are blown apart by a bad outcome" [15B, male].

Thus, for pregnant women, both doctors and peers and everyday companions act to reinforce certain norms of behavior. An obstetrician noted that "what often happens is that not only is the patient anxious, but their family, their significant other [are too]. So you've got all these people like *hounding* you about what you can or cannot do" [27B, female]. Women are subject to extreme and overt social control during pregnancy.[32] That social control reflects the general social interest and investment in new life, in the "next generation." "Protecting children is central to social life."[33]

In addition, drinking, like many other risky behaviors in pregnancy, is construed as a lifestyle choice that a woman *ought* to control. If only she had avoided "that one glass of wine," the child would have been unharmed. Abstaining from alcohol is seen as a test of a mother's fitness as a parent, indeed of her love for her child. "It sure warms my heart to see a woman stopping [drinking] totally because she values the intactness of the child she is bringing into the world. That's love," in the eyes of a pediatrician [10P, male]. A family practitioner likewise connected a woman's behaviors during pregnancy with the expression of maternal love: "So I think people have that mentality a little bit better in their heads, which is I am responsible for this baby in a very metabolic fashion. In addition, am I going to love this baby? And therefore I have to be very careful about what I eat, you know, as well as how much weight I gain. I mean, I think they lump all those together . . . because that's the care the baby is getting; it just happens before it is born, rather than after" [29F, female]. Reconstruing behavior during pregnancy as a test of a woman's love for her child — and indeed a test of her fitness as a mother — is indeed a powerful means to exert social control over women's actions. It is precisely because women are subject to such paralyzing guilt that exhortations not to drink while pregnant take on the force that they do. How can a pregnant woman justify to herself or to anyone around her having a drink, when to do so is to fail to love her child, to declare "my priority is not my fetus," as one pediatrician angrily put it? In some sense, a woman's behavior during pregnancy is regarded as a bellwether for her fitness as a mother. "The category of the 'pregnant addict' is thus saturated with social meaning . . . For those distressed by women's drift from 'selfless motherhood,' pregnant addicts represent women's refusal to postpone their momentary 'pleasure' (addiction) for the interests of the fetus."[34]

Doctors often saw women as having two responses to the diagnosis of FAS. Some women they saw as "so far gone in alcoholism, that nothing matters" or so "deep in their addiction" that they "really, truly don't get it." However, for other women the diagnosis has the power to induce severe guilt. "I think it would be devastating for a woman to learn that she has through her carelessness or indifference permanently damaged another, her own child," said a pediatrician. Likewise, an obstetrician noted, "Alternatively, [for] some of them that used to drink and then they have cleaned up their act and now they have got their lives together, knowing that they did something that harmed their kid, has a pretty significant impact." The threat of being held accountable for harm to one's own child can exert great force on women's behavior.

Maternal-Fetal Conflict in Pregnancy: "Don't Do This for Yourself, Do It for Your Fetus"

The diagnosis of FAS epitomizes the paradigm of maternal-fetal conflict that is typical of current medical and lay thinking about pregnancy. Many doctors in the sample were sensitive to this conflict and questioned "to what rights, no, to what *protections* is the fetus entitled?" They recognized the occasional tension between what they construed as a woman's freedoms or rights and the well-being of the fetus.

> I think that this society is deeply conflicted, because on the one hand there is the
> idea of a woman's right to privacy and a right to have a termination of pregnancy

if she wants it, up until some point in pregnancy, . . . and on the other hand . . . depending on one's ethical or religious views, that's either moral or immoral. But viewed purely in the temporal frame of reference of the rules of the road of a society, women's autonomy is one set of issues; and then the other set of issues are public health issues: at some point, what is the duty of the state to protect somebody from health hazards? And then the question is when does a fetus become a somebody? Because we as a society are going to be picking up the tab if that kid is in special ed, but does that give us the right to dictate to a mother, you know, if you . . . go into a bar, do you have to produce a urine-for-pregnancy test before they'll serve you a drink? [19P, male]

As this doctor expresses, the tensions inherent in this conflict extend beyond the woman and child to the broader society. In drinking during pregnancy, a woman not only threatens her child, she imposes costs on society as well: "we are going to be picking up the tab." Thus, maternal behavior during pregnancy is construed to have social as well as individual consequences.

Other doctors characterized the conflict as between a careless or irresponsible mother and a defenseless fetus. "If you put a toxin into your body, somebody is going to suffer," warned a pediatrician. The implication here, of course, is that it is the fetus, not the woman, who suffers. Some obstetricians objected to this displacement of concern from the woman onto the fetus: "What's happening is that people are saying don't do this for yourself; do it for your fetus." But in the eyes of these doctors, the woman "is not [just] a uterus . . . [Drinking] is a shame for the fetus, but it's a shame for this woman too" [28B, female]. They were troubled that women who suffer alcoholism and harm themselves through their drinking seem unworthy of attention until they become pregnant and their drinking threatens not only themselves but their child and, by extension, society. Obstetricians struggled to keep the focus on *their* patient, the woman. In fact, many obstetricians recognized that "pregnancy is a motivating factor," a time of life when women are especially concerned about acting healthfully. "I think that most women feel a tremendous sense of responsibility for what they do or don't do in the pregnancy . . . There have been very, very few women in my life that I have ever met that I truly believe that they didn't care. I think that's very, very rare" [15B, male]. Doctors hypothesized several reasons for the power of pregnancy to motivate women. They saw pregnancy as a "public," highly visible condition, in which women are subject to the kinds of overt social control mentioned earlier. One doctor thought we fo-

cus on pregnancy as much as we do because it is visible, because women's public behavior is easily monitored. Pregnancy is thus made into a "special state"
in which it is acceptable to violate women's privacy, to cross otherwise invio-
lable social boundaries. Pregnant women commonly experience this trans-
gression physically; strangers, including men, are likely to reach out and touch
or pat a pregnant woman's belly, in a gesture that would otherwise be considered a serious — indeed unthinkable — breach of common decency. "It's rude!"
protested a female obstetrician [27B]. That strangers feel so free to violate
norms demonstrates once again the ways in which pregnancy is public and
pregnant women are regarded as communal property.

Perhaps more important, though, most pregnant women are genuinely
motivated to have good outcomes. "When I think of what birth means kind of
sociologically to people, and you're trying to give your kid the best chance, and
you're always wanting your kid to do better than yourself and to really manage
in the world and make it, and this [not drinking] is something you can do to
help that" [1F, female]. More generally, doctors acknowledged that pregnancy
is a unique period in a woman's life. Doctors credited the unique relationship
between the woman and the fetus with motivating behavior change during
pregnancy. "I think there is more awareness that you are all one interrelated
unit. I think pregnancy used to be considered, you know, like a parasitic
growth. I don't mean that people really thought of it that way, but their perception was 'This is something on autopilot. And I am just holding onto it and
then I am going to return it to the library,'" speculated one family practitioner
[29F, female]. In the words of one of her colleagues,

> Culturally we say that pregnancy is the most — I don't know about the most — but
> one of the most valued assets that anyone can have. It's the only thing that, because
> that nine months, your contribution, the mother's contribution to that nine
> months, has a, can have a profound influence on a lifetime, and so culturally, we
> say that's just a supervaluable period of time . . . There is no other interaction:
> when they are in the womb, like, it's all there. Like, you know, there is a protec-
> tive environment, literally. Those external influences don't predetermine whether
> they are going to end up in trouble or not. Whereas once they're out, no matter
> what you do, those external influences become the real strike. [30F, male]

Thus, during pregnancy, the woman has the potential to shelter and protect
her fetus from the welter of external influences that will inevitably mark the

child's life. Pregnancy is "supervaluable" because it provides an opportunity to afford the unborn child a "perfect" environment, however briefly. Pregnancy affords an opportunity for control that will be irretrievably lost postpartum.

Doctors sometimes self-consciously used women's desires for a positive outcome to motivate them to change their behavior. "Would you be interested in stopping drinking, if not for yourself, for the baby?" asked one family practitioner of her pregnant patients who drank. "I think it's an easier sell to tell people not to drink during pregnancy . . . because it's not just for them. It's very clearly not just for them, so I think you end up able to get people to buy into that more easily," she contended [29F, female]. Moreover, such moral suasion is particularly powerful to women, who are socialized to value self-sacrifice.

However, doctors recognized the powerful subtext of this focus on health and well-being during pregnancy. Obstetricians in particular expressed concern that women's drinking elicits a medical response *only* when they are pregnant. One obstetrician lamented that if we would only focus on problem drinking throughout the lifecourse, we would not face the problem of drinking during pregnancy. To single out drinking during pregnancy as problematic seems shortsighted, she thought.

Regardless of specialty, most doctors saw pregnancy as a complex, even mysterious process, one that medicine is somewhat at a loss to understand. Not surprisingly, obstetricians were particularly attuned to the opacity of pregnancy, which in many ways symbolizes the uncertainty that inheres in all of medicine.[35] "The increased professional and public preoccupation with medical uncertainty notable throughout the 1960s and 1970s has been centered on problems of error and risk, hazard and harm, as well as probability and predictability."[36] Certainly the diagnosis of FAS fits this pattern. Doctors readily acknowledged that even medical understanding of pregnancy remains incomplete, that there are many things we simply do not know. "The biggest contributor to adverse birth outcome is the category of 'don't know,'" said an obstetrician. "Don't know why that patient — clearly it was nothing under the mother's control and she does not have any risk factors for anything bad, but yet has a horrible outcome whether it is an anomalous baby or a preterm birth. By far, the largest, the two largest categories for adverse birth outcome are anomalous babies and preterm birth, and for both of them, the largest cause is 'don't know,' so . . . can't do much about that" [14B, male]. "It's all pretty mysterious," agreed a family practitioner [18F, male]. Even pediatricians conceded that we "don't always know the exact causes of adverse birth outcomes in the kids that we see." In fact, the specific etiology of 75 percent of birth defects is unknown.[37] Thus, doctors acknowledged that, despite our rhetoric of personal responsibility, it often is not possible for them or for women to control the outcomes of pregnancies.

Doctors' Perceptions of Risk: "A Whole Complicated Milieu"

Given these somewhat conflicting cultural notions of risk and responsibility, and the inherent uncertainty of pregnancy, how do doctors evaluate risk in pregnancy? Although pregnancy is seen as an inherently risky state, the concept of reproductive risk has come under growing criticism. As the concept of high risk "became a nebulous idea that lost its power to discriminate among pregnancies, . . . [it] pervaded the space of obstetrical work."[38] Pregnancy itself was considered normal, but each individual pregnancy presented the occasion for risk of a potentially abnormal outcome. The concept of risk in pregnancy allowed obstetrics to extend the scope of its purview "farther back into the pregnancy, into a woman's history and her environment."[39] Thus, the notion of risk further medicalized pregnancy, subjecting a greater portion of the experience of reproduction to medical scrutiny and control. Several trends contributed to the ascendancy of the risk paradigm in obstetrics. Health more generally became an "epicenter of the increased malaise about uncertainty, and anxiety about danger and risk that have surfaced in our society."[40] Epidemiology increasingly focused on individual rather than ecological risk.[41] Finally, pregnancy itself was pathologized by obstetricians as a means of achieving professional monopoly.[42] One of the latent consequences of these trends was the diffusion of risk throughout pregnancy.[43] Everything that a woman did and experienced, everything about herself personally and about her environment more generally, became potentially risky. Moreover, "since the 1980s, assumptions of maternal vulnerability have been reconstructed around risks to the fetus mediated through the maternal body."[44]

Doctors in this sample did in fact see pregnancy as a "whole complicated milieu" in which women and their fetuses are subject to numerous and often intertwined risks. As one doctor put it, "Walking across the street could be risky for a pregnant patient . . . I could talk for two hours about risks and pregnancy" [4B, male]. Many doctors resisted the idea of ranking kinds of risk during pregnancy or of disaggregating one risk factor from its correlates. They saw risk — and risk-taking — as a package. "It's a whole mode," said one pediatrician. "I think there are a whole set of women who have things other than alcohol that are in the same group that, just, you know, no prenatal care, other exposures, drugs, that kind of stuff, so I don't know that it is the alcohol exactly, it's the whole mode" [20P, female]. Doctors saw these as "things that are interactive together." "I personally don't know how you can tease them [risk factors] apart," reported one doctor. A family practitioner said, "I don't rank them [risk factors] when they are all part of the same sort of what I consider to be risk-taking behavior. They are a cluster" [29F, female].

The intertwining of these risk factors also makes it difficult to pinpoint the risk posed by alcohol alone or to define the level of alcohol exposure that is risky. Many doctors saw alcohol acting in concert with other factors that might exacerbate its effect. "I think the whole notion [is] that there is environmental, there's nutritional, there is psychobiological things that influence your ability to say an exact amount [of alcohol that is risky] is the problem" [16F, male]. "All the literature on the effects of fetal toxins say[s] that it is extremely difficult to know what effects this particular one has because there are usually so many other things going on. To separate it out, say this is all alcohol, whereas this is — and to factor out smoking, the malnutrition which often accompanies, frequently, usually accompanies the alcoholism, and the effects of poverty, whatever they may be, although it is not just poor people who drink, it's very difficult to separate them out and rate them" [10P, male]. Several doctors noted that heavy alcohol use is usually accompanied by smoking. "There is no isolated use. Show me a patient who only uses alcohol and doesn't smoke. And that'll be the one patient that I've ever seen" [4B, male]. Others noted that "alcohol is a known appetite suppressant" and that patients who are alcoholics are frequently "into polysubstance use." Thus, the effects of alcohol on the fetus are inseparable from the effects of malnutrition, smoking, and other drug use.

Nevertheless, doctors saw "the things that people do as social habits" as most worrisome. "Honestly, it is probably better if you do heroin during pregnancy than if you drink, from the kid's standpoint . . . I think people sometimes confuse being socially unacceptable with being a worse teratogen, where that is not necessarily the case," said one obstetrician who specializes in maternal-fetal medicine [26B, male]. In fact, there is no evidence to date that either heroin or cocaine has specific teratogenic effects on the fetus. Like alcohol, cocaine use can cause intrauterine growth retardation. Perhaps even more seriously, cocaine use is a common cause of abruptio placentae (premature detachment of the placenta) and thus a major contributor to preterm labor and delivery in some populations. Some doctors reported that women seem to be aware of this relationship and sometimes take cocaine deliberately in order to go into labor and end their pregnancies early. (Interestingly, doctors also reported that some women view the growth retardation caused by cocaine use or cigarette smoking as positive, since they believe smaller babies will be easier to deliver.) Both cocaine and heroin may cause withdrawal symptoms in the newborn; more rarely, babies exposed to heavy doses of alcohol immediately prior to birth are reported to suffer delirium tremens and other symptoms of alcohol withdrawal. Doctors worried that patients are least concerned about behaviors that are most familiar to them, things like drinking and smoking that are a "dominant part of their lifestyle before they were pregnant."

Doctors varied as to whether they saw alcohol or smoking as more dangerous during pregnancy. Doctors commonly expressed the sentiment that "alcohol and cigarettes would be kind of one and two on my list. I'm not sure which would be one and [which would be] two." Some doctors weighted alcohol more heavily as a cause of adverse outcomes.

> If you have somebody who's a heavy smoker, as compared to a heavy drinker, probably the heavy drinker faces more prospect of long-term problems with the child that will be hard to take care of than the risks that are associated with the heavy smoking. The heavy smoking may be related to more miscarriages, smaller baby, but we haven't really associated it with *retardation* and *behavior problems*. And maybe I'm wrong about what the overall impact is, but it just seems to me that the lifelong character of what can happen with fetal alcohol is much more disruptive than what we know about for smoking. [1F, female]

However, other doctors found smoking during pregnancy more troubling than alcohol use. "I personally think that there should be more concern about smoking during pregnancy than drinking during pregnancy. Because we know that [for] every woman who smokes, there is a reduction in birthweight for the fetus. It's more prevalent, yet it is socially acceptable and it's a significant problem. I mean, I'm not discounting the fact that alcohol is a problem as well."

Although some doctors saw alcohol as "one of the best-described and most common teratogens . . . on the teratogenic hit list, it's probably number one" [24P, female], others were quick to assert that they were equally or more worried about other kinds of risks. "I'd like to think that I was equally worried about a lot of things," said one. Several doctors rattled off long lists of risks: "poor di-

ets, sexually risky behaviors, IV drugs, alcohol, smoking" or "low socioeconomic class, multiple sex partners, multiple drug use, smoking." "I would put
drinking number three [after] nutrition, smoking, in terms of intrauterine
growth retardation." Often they acknowledged that, in terms of absolute numbers, relatively few pregnant women drink, so that more common practices,
like smoking or poor nutrition, affect greater numbers of pregnancies.

> I am not most concerned about a woman doing that [drinking during pregnancy],
> because I think there is a small percentage of women doing that. But if you talk
> about modifiable things, about doing during pregnancy, I would say drinking is
> the most important that I can think of and smoking comes in second. Taking your
> prenatal vitamins probably comes in third, but the most important to me is taking
> your prenatal vitamins because that can help 100 percent of the population,
> whereas drinking, while I ask about it, only affects a smaller percentage of the pop-
> ulation. So I guess it depends whether you [focus on] impact of intervention on
> the individual or impact over the population . . . Like, public health–wise, drink-
> ing is not as important as prenatal vitamins, but individual health–wise, drinking
> is more important than prenatal vitamins, and I'd rather have somebody that didn't
> take their prenatal vitamins and didn't drink than someone who took their vita-
> mins and drank. [30F, male]

Most doctors, like this family practitioner, noted that alcohol use, although
rare, caused potentially more devastating results than other risk factors. "I don't
know that, I am not sure I know that alcohol consumption during pregnancy
has any more impact than tobacco or a lack of taking your folate[45] or again,
poor socioeconomic status as it relates to a good follow-up or prenatal care. I
don't know that it is any more important than any of those other factors" [11P,
male].

However, there is one important way in which doctors singled out drinking from other risk factors during pregnancy. They regarded drinking as a "preventable" or "modifiable" risk. Doctors referred to alcohol's effect on an "otherwise healthy pregnancy." They also saw alcohol as particularly harmful in the
permanence of its effects. The effect of cocaine "often washes out. But, boy, if
you've got FAS, you've got it. It's with you . . . these changes are definitely permanent and therefore really worrisome" [10P, male].

Managing Risk in Pregnancy: "There Are So Many Things That You Can't Prevent, and These Things You Can"

Fetal alcohol syndrome is widely cited as the leading preventable cause of birth defects in the United States. Physicians for the most part concurred with this assessment; many did not hesitate to say that FAS is preventable, "absolutely preventable." Some even seemed taken aback by the question, as though it were too obvious to answer. Many doctors felt a special urgency about preventing FAS, since "there are so many things that can affect a pregnancy that we don't know about that those that we do know about we can at least control for." They saw themselves as helpless to prevent many of the adverse outcomes patients could experience, particularly in terms of birth defects, which they saw as "unpreventable." In contrast, FAS seems to be an easy target: there is nothing ambiguous or mysterious about the relationship between alcohol exposure and this particular constellation of birth defects. As one geneticist put it, "there are so many things that you can't prevent, like an extra chromosome, and the child [born with FAS] was programmed to be normal and the fact that there was exposure caused this to happen" [20P, female]. "That's why, as a family physician, we tend to concentrate on things that are fixable and preventable and there are so few things in medicine that are," contended a family practitioner [29F, female]. Thus, doctors' beliefs that FAS is "absolutely preventable" hinged on their faith that this "exposure" could be eliminated easily — "this is such a simple thing to have avoided," whereas other kinds of problems in pregnancy are much more difficult to address. "A lot of that stuff is pretty hard to prevent short of a really big attempt to ensure good nutrition and good prenatal care and all that stuff. So I guess upon thinking about it, what are the preventable — and a lot of the birth defects are not preventable in any case — but FAS would be preventable. So it is probably one of the few really catastrophic pregnancy outcomes that could be prevented by a clear intervention: stopping drinking" [18F, male]. From this perspective, preventing FAS seems like a quick fix, something that is "controllable." "There is very little else that is preventable," so FAS seems particularly egregious, in a category of its own.

Yet some physicians saw a more complicated picture. They described FAS as only "potentially preventable" or "theoretically preventable." They answered conditionally, noting that "It's preventable if the ladies don't drink,

[and] that's a huge if." "Yeah, it's preventable," said a family practitioner. "Don't drink, don't get it. But if you say, if somebody drinks, is it preventable? No, I don't think so" [30F, male]. In other words, "stopping drinking" did not appear to be a "clear intervention" in their eyes. "What has not been shown is that there is an effective way to get people to stop drinking," conceded one obstetrician [21B, male]. If preventing FAS requires preventing "alcoholism in the mothers," doctors saw it as no easy task. A few answered the question "Is FAS preventable?" by asking in reply, "Is alcoholism preventable? Yes and no." Thus they saw FAS as part and parcel of a much larger social problem, one that is not readily amenable to easy solutions or even prevention. "I think that is a challenge. I mean, it would be nice to eliminate FAS, but it's a challenging proposition. Sure, it would be nice to be able to have alcoholic women not drink during their pregnancy, but other than locking them up somewhere when they are pregnant, how do you prevent alcoholics from drinking? I don't know the answer to that" [24P, female].

Whether doctors regarded FAS as preventable or not depended at least in part on their beliefs about why women drink. Doctors who believed that FAS is preventable believed that a woman's willpower and motivation are essential. Few doctors were entirely dismissive of the reasons women drink. Many agreed that alcoholism or drinking is an addiction, "it's probably something they have very little control over," "a medical problem" that requires more than willpower to overcome. In the eyes of one family practitioner, pregnant women drink "For the same reason most people drink. I think because they probably have emotional problems they are trying to make better. Socioeconomic things that are bothering them, depression, whatever. And they're trying to get relief" [13F, male].

However, some doctors problematized drinking even further. They saw the "roots and drivers" of addictive behavior and believed that women drink because of the deep troubles in their lives. "A lot of women . . . are in very horrible social situations while they're pregnant . . . I think that life is hard, and harder for some than others," acknowledged a pediatrician [12P, female]. Moreover, these doctors connected those personal troubles to wider social inequities. "The roots and drivers of it . . . are, you know, connected to capitalism and poverty and racism and, you know, things which are deeply woven into our culture really," speculated a family practitioner [18F, male]. One obstetrician observed, "I think there is a tension between whether something like drinking during pregnancy is a medical or a social problem. It can have a lot of social causes" [28B, female]. These more sensitive physicians saw social conditions as, in Link and Phelan's term, "fundamental causes"[46] of FAS. Family practitioners, in particular, were sensitive to this issue.

Some doctors, then, acknowledged that prevention rests on more than women and women's behavior alone. "But it means that as we try to tackle the issue of alcoholism with people, I think it immediately kind of gets you into this larger area of what it is that's giving meaning to their lives or why is there an absence of meaning to their lives, that they're feeling like this, and what kind of pain is that they're trying to anesthetize with the substance and addictive behavior" [18F, male]. Particularly as women's drinking behavior stems from the difficulties of their own lives, doctors saw it as difficult for them to change. "I think there is a large group of women who have some knowledge of certain things, but really don't have the ability to change what they do." Women who drink, particularly women who are unable to stop drinking even during pregnancy, are "feeling depressed, they want to get out of their — they want to escape from their current environment" [4B, male]. In the eyes of most doctors, drinking during pregnancy indicates "a significant amount of social distress" [5P, male].

Doctors also hesitated to blame women who drink during pregnancy. "You really have to back off [from blame] and appreciate that there is a lot more going on in that person's life that has overwhelmed her to the point that she has difficulty in fulfilling that responsibility, despite the fact that she probably wants to or does clearly have a concern," noted one obstetrician [15B, male]. "Either [they] don't have the knowledge, don't have the resources to change, are addicted, or are in such dire circumstances that they can't be concerned beyond the next immediate thing that they are going to do," said a pediatrician [9P, female]. "To put it simply," said one family practitioner, "I think [for] any addiction behavior, the real barrier to stopping is the psychological and sociological costs that it puts upon them" [30F, male].

Not all doctors had as deep an understanding of the complexities of women's drinking behaviors. Some disputed whether drinking is in fact a medical problem. One obstetrician noted, "I think that a lot of physicians don't believe it's a medical problem. They think it is a social problem. It's a behavioral problem, that people choose this addictive lifestyle and that they can on their own reform" [28B, female]. Some dismissed drinking as one of "many social work issues" that doctors are likely to see in clinic patients in particular. Others thought women drink because "It's part of your life, you're used to having a couple of glasses of wine with your dinner. How can you perceive that as being bad? . . . So it is very socially acceptable and that's why people have an easier time of doing it."

Alcohol as a Social Problem: "But the Problem Is Why Are People Drinking?"

Clearly, doctors perceived a much larger problem than drinking during pregnancy. They saw alcohol use in general as a much broader social problem than alcohol use in pregnancy. Regardless of specialty, doctors believed that greater attention ought to be given to the myriad social problems caused by alcohol use, that addressing these problems is of greater concern and priority than addressing drinking during pregnancy per se. "Oh God, we should do more about drinking, period." "It's just a social evil to me," concluded one pediatrician. "I don't know what the purpose of it [alcohol] is." Perceiving that drinking during pregnancy is only part of a larger problem, doctors were forced to confront the issue of priorities: which problem should we focus our attention on, drinking during pregnancy or drinking?

> We compartmentalize all those things. I think we should be looking at . . . why are they using alcohol? Because they have got problems within their lives. And until we resolve a lot of those problems and different reasons for the problems, socioeconomic, or spousal abuse, or whatever else is going on, people are going to go to drugs or alcohol to resolve their problems. So the woman who drinks in excess during pregnancy is, as I said, she has got a life that's overwhelming her. So until you deal with that life that is overwhelming her, you can give her literature up the wazoo . . . But we aren't addressing the causes. The causes largely are socioeconomic. [15B, male]

Doctors also argued that there are other more pressing problems that deserve attention. "There are so many women's issues that we should be addressing. We just don't have the time to do it or investigate it." "There is so much else to — so many other problems," worried one family practitioner.

> We should do, you know — there's many women who don't get any prenatal care, so do we put those resources to those who drink and are pregnant or those who don't get any prenatal care and deliver tiny babies? I don't know. [16F, male]

I think that if we are going to do more, it should dovetail with trying to improve their nutrition. It goes with trying to improve people's lives when they are pregnant. Not just get them to stop drinking, but I think the problem has more to do with the mental health system . . . because I think if these people were involved in the mental health system, we wouldn't have so much drinking in pregnancy . . . You know, in terms of the whole issue of poverty and lifting people out of poverty. I mean it all goes along with that. [28B, female]

Moreover, many doctors saw FAS not just as a somatic disorder in the child but as an indicator of social disorder within the life of a woman and her family. "They're babies with a problem . . . I mean there are lots of problems that babies and children have. And they're just kids with problems, I mean, so you work with them . . . and deal with whatever the parent's problem is . . . I don't really perceive fetal alcohol to be, I mean, I know it is a specific thing, but in some ways, it's also generic . . . In many ways, my emphasis is on trying to help the parent, so that the kid can be helped . . . I see it as another situation where the more help you get for the parent, the better off the kid'll be" [1F, female]. Thus, for family physicians, FAS was clearly a diagnosis not just of the child but of the mother as well, and the family practitioner saw both mother and child, the family unit, as the patient in need of medical attention. It is the "problems" in the family that require treatment rather than the syndrome itself. In that sense, FAS is "generic," merely an indicator of a set of complex social instabilities and risks.

Just as a child born with FAS is an indicator of a compromised parenting situation, so drinking during pregnancy is a symptom of deeper, more pervasive troubles in a woman's life — and indeed in society at large. The real problem is not drinking during pregnancy but the woman's substance abuse itself.

The more important issue is why is she drinking? And you've got to deal with that issue because that lady's life is being blown apart. Can you blame her because her life is being blown apart and she can't handle that, so she deals with it in the form of some type of alcohol? My wife jokes about the fact that we live out in Bryn Mawr, on the Main Line,[47] and how the Main Line women for years have dealt with their issues of life, not pregnancy-related, but issues of life, by making sure that their martinis were strong. If the alcohol levels stayed high enough, hey, it's not a problem. So what's different from that than the lady in the ghetto? Her life is a hell of a lot tougher. [15B, male]

What requires urgent attention is the forces that "blow lives apart" rather than the drinking, which is only a symptom of such devastation. "I'd have to couch it in larger terms in that I think that we as a society need to think more about addiction as an issue in our culture." Yet doctors conceded for the most part that substance use is "something we have no control over."

How Medicine Confronts Social Problems: "It Is Our Job to Do That"

However perceptive their recognition that both drinking during pregnancy and FAS are indicators of deeper troubles in their patients' lives and in society more generally, physicians admitted they felt inadequate and unprepared to confront these social problems. "I think I would probably say that the medical community as a whole has been stymied in its attempts to deal with addiction issues in general and alcoholism in particular, and that extends to the area of obstetrical care . . . I mean, I think it is part of our job to do that, but I think it is a long run for a short gain and it is a pretty hard activity to do" [18F, male]. Like this family practitioner, many doctors reported that they found dealing with addictions in their patients difficult. They conceded that their profession "doesn't deal well with substance abuse issues." An obstetrician said, "It's very hard to find people to treat . . . any patient that has a problem with addiction" [28B, female]. Fetal alcohol syndrome forces doctors to confront, sometimes painfully, the limits of medicine. Many acknowledged that "it's a hard topic, very hard." Doctors speculated that their discomfort dealing with substance abuse in their patients springs in part from embarrassment and reluctance to confront patients. One obstetrician said of his colleagues, "Maybe they don't know the right questions, maybe they are afraid their patients are not going to like them because they are asking these things" [21B, male]. They saw themselves as "not well trained" to deal with these issues. "Obstetricians and family practitioners and pediatricians are the ones who should be — and internists — I mean, they should all be raising it with their female patients, and I don't think that most of them do," complained one pediatrician [19P, male]. "Physicians tend not to want to ask their patients about alcohol and drugs, especially in the private sector," admitted an obstetrician. Doctors also shied away from dealing with patients' substance abuse problems because they saw such problems as overflowing the boundaries of a typical office consultation. "I think a lot of us don't get into it because once you open that door up, it is such a time-consuming phenomenon during the visit . . . You just give them the general spiel of 'don't drink in pregnancy' and of course that takes care of the issue" [15B, male].

Doctors felt a responsibility to help their patients, even as they admitted their ineptitude in doing so. "It's a disease that can affect her baby long-term and so you owe it to her to discuss it and take issue with it every single time you see her," said one obstetrician [21B, male]. Some doctors expressed guilt or blamed themselves for not being able to protect their patients from misfortune. One obstetrician said, "Number one, when you see an adverse outcome, the first feeling is what did I do wrong that could have prevented this? Or how could I have prevented this? So you are immediately looking for that: what, how did this happen?" [15B, male]. This doctor's confession suggests that obstetricians at least identify with their patients in important ways. In some sense, they not only empathize with the burden of guilt felt by the mother but also assume that burden as their own. Doing so makes it much less likely that these physicians will blame the woman for an adverse outcome; at the very least, they share the onus of guilt with their patients. Just as women assume a heavy responsibility for the outcomes of their pregnancies, some doctors too felt a deep obligation to help their patients achieve optimal outcomes.

To that end, doctors perceived a two-fold responsibility: to "raise the issue" of drinking with patients and to educate patients about risk. "We should make sure that pregnant women know that this is very risky business and that there is no known safe amount, and that should be broadcast from the rooftops" [10P, male]. Yet doctors also must temper their warnings to patients so as not to induce excessive guilt. "I tend to get the message across, but not beat her over the head with it." "You don't want to make your patients really freak out over this as well . . . So I'm not, I'm sort of not hellfire-and-brimstone about it. I just try to explain to the patient the best that I can what the risks are" [21B, male].

Thus, while many doctors have a clear sense of the larger patterns of social inequality and suffering that cause drinking during pregnancy and FAS, as clinicians they nonetheless deal with individual patients, one at a time. Their interventions take place at an individual rather than a social level. Doctors see themselves and their colleagues as responsible for treating individual patients, not for solving social problems. "I am not a policymaker," objected one obstetrician.

One pediatrician raised the issue of how doctors situate their own responsibility towards their patients. "I have no idea what obstetricians say to women. I *hope* that they are campaigning strenuously against any drinking at all . . . I feel that obstetricians are mostly concerned about the women and not always as much concerned about the infants or fetuses as they should be" [10P, male].

This pediatrician's comment highlights a leitmotif of this chapter: the type of patient doctors feel responsible for — woman, fetus, infant, child, or family — influences how they think about the risks posed by drinking during pregnancy. The "knowledge that guides conduct in everyday life"[48] varies somewhat among the specialties. Obstetricians seemed as likely to be concerned about the harm an alcoholic woman inflicts on herself as about the harm to the fetus; they were even occasionally frustrated by the focus on the child to the exclusion of the woman herself. Pediatricians, on the other hand, tended to be far more protective of children. These differences are evident not just in the positions that doctors take with regard to risk but even in the language they use. Family practitioners and pediatricians, for example, talked about "babies." Obstetricians more commonly spoke of the "fetus." Pediatricians were more likely to see the affected child as a "blameless victim"[49] or, in the words of one of the pediatricians in the sample, "a child who has no voice," but obstetricians and many family practitioners were inclined to see the woman herself as a victim. These differences are in large part attributable to the very different (albeit interconnected) patients that obstetricians, pediatricians, and family practitioners feel responsible for. Doctors' perceptions of risk and of responsibility — both their own and their patients' — varied according to their specialty. "Who is your point of focus for those risks? Is it the fetus or is it the mother or is it the family unit?" asked one obstetrician. Another acknowledged, "It's funny — who your patient is is your main concern." Other doctors spoke of a "dual responsibility" to the mother and to her child, as well as to future children she may conceive. One pediatrician saw it as her responsibility to warn "a young woman that's going to still have other children, there has to be — I think you have an onus as a physician that she is totally aware of what has happened, but again feels somewhat empowered to move on from that point" to prevent other children from being born with FAS. Physicians thus assumed a broad spectrum of responsibility. Moreover, regardless of their specialty, most doctors saw alcoholism, if not alcohol use, as a wider, more pressing, and more serious social problem than drinking during pregnancy per se.

That I did not see greater differences by gender suggests the extent to which professional socialization overwhelms individual gendered understandings of the world. Among the obstetricians, this absence of gender difference is also a testament to how the men were aware of and able to empathize with female views and problems. In contrast to prevailing stereotypes about male obstetrician-gynecologists (such as those portrayed by Scully),[50] no male doctor in my sample was overtly misogynistic.

As physicians struggled to make sense of the diagnosis of fetal alcohol syndrome, the "raw confusion of the real world"[51] embodied in their patients, be they women, infants, or children, continuously impinged on their consciousness and their construction of FAS. In fact, the diagnosis is neither easy to make nor easy to explain, much less easy to prevent or to treat. Gusfield claims,

> As ideas and consciousness, public problems have a structure which involves both a cognitive and a moral dimension. The cognitive side consists in beliefs about the facticity of the situation and events comprising the problem — our theories and empirical beliefs about poverty, mental disorder, alcoholism, and so forth. The moral side is that which enables the situation to be viewed as painful, ignoble, immoral. It is what makes alteration or eradication desirable or continuation valuable.[52]

Doctors' ideas about FAS encompassed both cognitive and moral elements. Certainly doctors associated certain "facts" with the diagnosis: lists of symptoms, periods of exposure and levels of drinking, sets of interrelated risk factors. These facts were often expressed in clinical and technical terms — doctors spoke of teratogenic effects, microcephaly, microsomia, micrognathia, intrauterine growth retardation, organogenesis, and ethanol. But doctors' conceptions of what the diagnosis of fetal alcohol syndrome means went much deeper than the official language of their profession allowed them to convey. They also superimposed their own understandings of how the world works, of fairness and suffering, of fortune and misfortune, of human relationships, duties, and obligations, as well as their own moral judgments — whether permissive attitudes towards drinking during pregnancy or strict prohibitions against it — on their official understandings of the disorder. Thus, a doctor's personal beliefs became codified as clinical knowledge. As one doctor told me, "When you see it [FAS] in your practice, it's real. It's the truth." Ultimately, experience confirms reality.

Finally, although doctors universally saw fetal alcohol syndrome as "a big tragedy," they located the origins and victims of that tragedy differently. Some saw tragedy in the children who suffer the disorder. Others saw tragedy in the "blown apart" lives of the women who drink during pregnancy. Still others located tragedy deep within our society itself, "in the great emptiness in our culture," in the forces of inequality and suffering, of social disorder, that marked women with addiction and children with a physical disorder unamenable to medical intervention. In doctors' eyes, fetal alcohol syndrome and drinking during pregnancy are "horribly sad" indeed.

CHAPTER 5

DISCORDANT DEPICTIONS
OF RISK

All statistics are the products of social processes. Someone has
to decide what to count and how to go about counting.
—Joel Best

As we saw in Chapter 4, doctors vary in their assessments of the extent of drinking during pregnancy. Just how many infants with fetal alcohol syndrome are born in the United States every year? Although this question might seem an easy one to answer, defining the incidence of FAS is in fact a difficult matter. Even sources that might be regarded as definitive — the Institute of Medicine of the National Academy of Sciences and the federal Centers for Disease Control and Prevention — disagree on the number of births affected by FAS. Moreover, their estimates are not precise but span a wide range: the IOM estimates that 2,000 to 12,000 children with FAS are born every year; the CDC puts the number between 1,300 and 8,800. Fetal alcohol syndrome is often identified as the "leading cause of preventable birth defects," and drinking during pregnancy is construed as a major public health problem that affects a large number of babies every year. But the incidence of FAS is not as clear as these claims suggest. We do not know how many babies are born with fetal alcohol syndrome every year. Nor do we know how prevalent drinking during pregnancy is or which women and children are most likely to be affected by fetal alcohol syndrome.

If we consider the official estimates from the IOM and the CDC, as few as 1,300 or as many as 12,000 children are born with FAS every year. Certainly these numbers are not trivial, but neither do they indicate an epidemic of alcohol-affected births. To put these figures in perspective, consider that there are just over 4 million births in the United States annually. Some 300,000 babies are born with a low birthweight. About 150,000 are born with birth defects, approximately 1 in 33 babies. Cerebral palsy affects about 10,000 infants every year. Down's syndrome, one of the most recognizable birth defects, is diagnosed in about 4,000 newborns a year, and there is about the same number of neural tube defects, including spina bifida and anencephaly. Cystic fibrosis afflicts about 1,600 newborns each year. Finally, about 400,000 infants are born every year to mothers who smoked during pregnancy; these infants are likely to suffer growth retardation and other adverse effects. Thus, even the highest estimates of FAS are less than the incidence of other, more common adverse birth outcomes.

This chapter begins with a review of what we know about the epidemiology of FAS, the difficulties and uncertainties about the diagnosis itself, and the problems associated with determining who has FAS. I then propose an alternative measure of the distribution of risk, using data from the 1988 National Maternal and Infant Health Survey. To understand the dimensions of the problem of FAS, we need to know the answers to two questions. First, which women are at risk of drinking during pregnancy? Second, which women are most at risk of adverse outcomes from prenatal drinking? This chapter outlines the parameters of prenatal drinking in order to gauge how well the rhetoric of risk in pregnancy corresponds with the reality.

The Difficulty of Measuring FAS

Fetal alcohol syndrome presents particular measurement challenges. A 1996 report on FAS from the Institute of Medicine notes, "The boundaries of the diagnosis, as well as the markers that should be used to delineate those boundaries, are perhaps the most vexing issues."[1] Thus, at least two measurement dilemmas are inherent in the diagnosis of FAS. The first is what to measure. Should only so-called "full-blown cases" of FAS be included? Or should we also include the incidence of "fetal alcohol effects," "alcohol-related birth defects," "partial FAS," or any of the other designations that have been proposed to describe the spectrum of effects attributed to alcohol exposure in utero? What counts as FAS and what is "merely" FAE? In other words, what are the boundaries of the problem? The second dilemma is how to measure FAS in the population. Since there is no "gold standard" for the diagnosis, how do we

identify cases of the syndrome? These two problems present the deepest challenges to estimating the incidence of FAS, but we might add a third, which
may be summarized as when and where to measure it. Should we look for FAS
only at birth? Throughout infancy? Throughout childhood, throughout life?
And should we try to measure FAS at the population level? By registry-based
studies? In clinic populations? In particular racial or ethnic groups?

Many factors contribute to the difficulty of measuring the incidence of the
syndrome. Principal among these is the difficulty of making the diagnosis itself. First, there is no biological marker for the syndrome and thus no definitive laboratory test. There is no gold standard for the diagnosis and, in fact,
"none of the abnormalities found in FAS is specific to that diagnosis."[2] Thus,
FAS remains a diagnosis of exclusion. In the lingo of physicians, FAS is a "rule-
out"; that is, clinicians rule out all other possibilities before they decide a child
has FAS. It is the province of trained dysmorphologists and clinical geneticists,
and even these specialists can disagree about whether or not a patient has FAS.[3]
"In the absence of accurate, precise, and unbiased methods for measuring and
recording the severity of exposure and outcome in individual patients, diagnoses will continue to vary widely from clinic to clinic," acknowledged a recent article from the Dysmorphology Unit at the University of Washington
School of Medicine in Seattle.[4]

Second, the diagnosis is impressionistic and highly subjective. In fact, doctors struggle to describe the diagnosis succinctly, and they offer a range of
descriptions and definitions of FAS. They describe it variously as a "complex
syndrome," a "spectrum" of effects, a "constellation" of findings, an "impressionistic diagnosis," or a "diagnosis of exclusion" — or even "a trash-basket diagnosis," meaning that doctors "dump" otherwise unclassifiable children into this
catch-all diagnosis. In my interview sample, one obstetrician who specialized in
maternal-fetal medicine was skeptical about both the validity and the reliability
of the diagnosis. "The definition is such a moving target . . . There are those
who really get on the bandwagon — you know, if your only tool is a hammer,
the whole world starts to look like a nail. You take a baby to someone who is a
fetal alcohol expert and they're much more likely to come up with something
to resemble that [diagnosis]. Depends on who's making it [the diagnosis]. It's so
subjective. There are very limited objective criteria. As far as I can tell, they are
all subjective. I can't even think of an objective criteria" [4B, male]. In other
words, fetal alcohol syndrome is in the eye of the beholder, and the doctors who

make careers out of the syndrome are more likely to see it in children than are their colleagues. It is subject to ascertainment bias and therefore may be over-diagnosed in certain populations and underdiagnosed in others.[5]

In an effort to reduce this variability in diagnosis, the Seattle FAS group has developed two diagnostic tools. One is a numerical coding system that uses "quantitative, objective measurement scales" to determine the magnitude of each of the four components of the diagnosis of FAS: growth deficit, facial phe-notype, central nervous system dysfunction, and prenatal alcohol exposure.[6] The other is a computer program, trademarked as the FAS Tutor CD-ROM, which assists doctors in diagnosing FAS. The program uses an algorithm to de-cide whether or not a patient has FAS based on certain criteria that the doctor measures, including the length of the philtrum (the groove between nose and upper lip) and the thinness of the vermilion (the colored part of the upper lip), ranked on a scale.[7] Such a program is an attempt to remove observer bias from the diagnosis; it is a kind of false promise of the objectivity that is supposed to come from quantification. That such a program exists at all, however, only re-inforces the subjectivity, uncertainty, and elusiveness of the diagnosis of FAS. There have also been attempts in recent years to facilitate the ability of ordi-nary clinicians to recognize and diagnose FAS. These include developing clin-ical checklists to aid in the recognition of FAS and exploring the use of MRI or PET to diagnose FAS. Even more incredibly, one researcher has suggested that "The brain is the organ most sensitive to the prenatal damage caused by alcohol. If that damage resulted in a unique and unmistakable psychological profile, FAS could be diagnosed reliably without reliance on the variable and subtle clinical features."[8] To find "a unique and unmistakable psychological profile" without variation or subtlety seems ambitious indeed.

Third, the syndrome is difficult to diagnose at birth, for a variety of rea-sons. The craniofacial anomalies can be difficult to detect in the newborn, es-pecially when the head and face are temporarily misshapen after passing through the birth canal. Intrauterine growth retardation is not specific to FAS and may be caused by a host of other factors; moreover, some alcohol-affected babies may be born with a normal birthweight and only later experience growth retardation. The central nervous system disorders and mental retarda-tion that are symptomatic of FAS are not apparent at birth.[9] Thus, most mea-sures of FAS at birth are thought to give underestimates of the syndrome.[10] In fact, one study estimated that 89 percent of cases of FAS were diagnosed after the child was six years old.[11]

Fourth, both the severity and the type of symptoms vary. Not all children with FAS exhibit all the symptoms associated with the syndrome. Moreover, the expression of alcohol-caused fetal deformities can range from mild to severe: there is a wide range of phenotypic responses to in utero exposure to alcohol.[12]

Fifth, the diagnosis may be complicated by the patient's age and race. There are age-specific variations in the expression of the phenotype; as children age, their distinctive FAS facies may become less apparent, and they may experience catch-up growth and lose the small stature considered indicative of FAS. And recognition of the facial characteristics can be complicated by race. For example, a moderate degree of midfacial hypoplasia is a normal feature in Native American groups but may be easily mistaken as a sign of fetal alcohol syndrome.[13] This variability and confounding of the FAS phenotype seriously complicates what is already a tricky diagnosis. Moreover, the confusion between features of racial phenotype and features of FAS may contribute to overdiagnosis of the syndrome in certain populations.

> In summary, the current criteria for the diagnosis of FAS depend on recognition of a consistent pattern of minor, often subtle physical anomalies; generalized but disproportionate growth retardation; and non-specific developmental and behavioral aberration. Some of these characteristics change with time, and their degree of severity may vary among individuals. Confirming the diagnosis of FAS in a specific patient is often difficult, even for clinicians with considerable experience with FAS.[14]

Birth certificates were revised nationally in 1989 to include a standardized set of check-off boxes for reporting FAS and other congenital anomalies. Curiously, a CDC study in Georgia found that, overall, birth certificates both underreported and overreported cases of FAS. In other words, they had high rates of both false negatives and false positives. In fact, 71 percent of the cases of FAS identified on birth certificates were deemed to be false positives. The authors of the study speculate that the high rate of false positives "represents an overreporting of FAS without indication of physical findings from medical records to substantiate the report. Maternal history of alcohol consumption during pregnancy may be considered sufficient evidence by some health professionals for a diagnosis of FAS on the birth certificate."[15] These diagnostic pitfalls create a very real conundrum of measurement: how do we measure at the pop-

Table 5. Estimates of the Incidence of FAS in the United States

Source	Population and Time Period	Cases per 10,000 Births
CDC Birth Defects	United States, 1981–86	
Monitoring Program	White	0.9
(BDMP)[a]	Black	6.0
	American Indian	29.9
	Asian	0.3
	Hispanic	0.8
Population-based	American Indians, 1980–82	
screening[b]	Navajo	14.0
	Pueblo	20.0
	Southwest Plains	98.0
Expanded birth certificate[c]	United States, 1989	2.0
National Birth Defects Prevention Network[d]	16 states	0.4–4.9
Multiple data surveillance, Alaska[e]	Alaska Natives, 1978–91	21.0
Metropolitan Atlanta Congenital Defects Program (MACDP)[f]	Atlanta, 1989–92	2.3
CDC BDMP[g]	United States, 1979	1.0
CDC BDMP[g]	United States, 1992	3.7
CDC BDMP[h]	United States, 1993	6.7
California BDMP[i]	California, 1983–88	
	White	1.0
	Black	5.8
MACDP[j]	Metropolitan Atlanta, 1983–88	
	White	0.2
	Black	3.7
Fetal Alcohol Syndrome Surveillance Network[k]	Alaska, 1995–97	15.0
	Arizona, 1995–97	3.0
	Colorado, 1995–97	3.0
	New York, 1995–97	4.0

[a]Gilberto F. Chavez, José F. Cordero, and José E. Becerra, "Leading major congenital malformations among minority groups in the United States, 1981–1986," *Morbidity and Mortality Weekly Report* 37, no. SS-3 (1 Jul. 1988): 17–24.

[b]Philip A. May, Karen J. Hymbaugh, Jon M. Aase, and Jonathan M. Samet, "Epidemiology of fetal alcohol syndrome among American Indians of the Southwest," *Social Biology* 30, no. 4 (1983): 374–87.

[c]Stephanie J. Ventura, "Advance report of new data from the 1989 birth certificate," *Monthly Vital Statistics Report* 40, no. 12, suppl. (1992).

[d]National Birth Defects Prevention Network, "Congenital malformations surveillance report," *Teratology* 56 (1997): 115–72.

continued

Table 5 (continued)

[e]Centers for Disease Control, "Linking multiple data sources in fetal alcohol syndrome sur-veillance—Alaska," *Morbidity and Mortality Weekly Report* 42, no. 16 (30 Apr. 1993): 312–14.

[f]Centers for Disease Control, "Birth certificate as a source for fetal alcohol syndrome case ascertainment——Georgia, 1989–1992," *Morbidity and Mortality Weekly Report* 44, no. 13 (7 Apr. 1995): 251–53.

[g]Centers for Disease Control, "Fetal alcohol syndrome—United States, 1979–1992," *Morbidity and Mortality Weekly Report* 42, no. 17 (7 May 1993): 339–41.

[h]Centers for Disease Control, "Update: trends in fetal alcohol syndrome—United States, 1979–1993," *Morbidity and Mortality Weekly Report* 44, no. 13 (7 Apr. 1995): 249–51.

[i]Jane Schulman, Larry D. Edmonds, Anne B. McClearn, Nancy Jensvold, and Gary M. Shaw, "Surveillance for and comparison of birth defect prevalences in two geographic areas—United States, 1983–88," *Morbidity and Mortality Weekly Report* 42, no. SS-1 (19 Mar. 1993): 1–6.

[j]Centers for Disease Control, "Surveillance for fetal alcohol syndrome using multiple sources—Atlanta, Georgia, 1981–1989," *Morbidity and Mortality Weekly Report* 46, no. 47 (28 Nov. 1997): 1118–20.

[k]Centers for Disease Control, "Fetal alcohol syndrome—Alaska, Arizona, Colorado, and New York, 1995–1997," *Morbidity and Mortality Weekly Report* 51, no. 20 (24 May 2002): 433–45.

ulation level something we are not even sure how to recognize at the individual level? And how can we judge fetal alcohol syndrome to be a "big" problem when we are not even sure how often it occurs in the population?

Current Evidence on the Incidence of FAS

It is not only the IOM and the CDC that have different estimates of the incidence of FAS. Estimates in the medical literature and in the popular press vary widely, sometimes by an order of magnitude. Table 5 summarizes estimates of the incidence of FAS in the medical literature. A rate of 10 to 30 per 10,000 births is commonly cited throughout the literature, but the national Birth Defects Monitoring Program of the CDC identified only 1,782 cases in 9,057,624 births from 1979 to 1992, a rate of 2.0 per 10,000 births.[16] Data from the expanded birth certificate form likewise yield an estimate of 611 FAS cases in a total of 3,665,991 births for 1989, a rate of 2.0 per 10,000 births.[17] It is likely that the BDMP and the birth certificate data underestimate the true incidence of FAS, since both data sets rely on diagnoses made during the early neonatal period when the facial stigmata of FAS may be difficult to detect and central nervous system disorders may not yet be readily apparent. Moreover, some alcohol-affected infants may have a normal birthweight.

Because FAS is so difficult to measure at the level of the national population, other studies have attempted to obtain more accurate estimates of the incidence of FAS by using multiple surveillance methods at the state or local level. Linking birth and death certificates, Medicaid claims, Indian Health Service case files, and private pediatric case files yielded an estimated prevalence of 21 cases per 10,000 births among Alaska Natives from 1978 to 1991. Birth certificate data alone identified only 15 percent of these cases.[18] The Metropolitan Atlanta Congenital Defects Program (MACDP), which also uses multiple sources to identify cases of FAS, found an observed prevalence of full FAS of 1.0 cases per 10,000 births for 1981–89 and estimated a prevalence of 5.1 cases of full and partial FAS per 10,000 births over this period.[19] For 1989–92, the MACDP identified a rate of 2.3 cases of FAS per 10,000 births.[20]

Smaller studies, measuring the syndrome in certain subpopulations, particularly American Indians, have also yielded varying estimates. For example, one study of FAS among American Indians of the Southwest identified a prevalence in children from birth to age fourteen of 16 cases per 10,000 among Navajos, 22 per 10,000 among Pueblos, and 107 per 10,000 among Southwest Plains Indians (primarily Apaches and Utes).[21] A similar study of Indians in the Northern Plains reported an observed incidence of 39 cases of FAS per 10,000 and an estimated rate of 85 per 10,000 births.[22] Such studies typically rely on screening of populations of schoolchildren by dysmorphologists. Studies like these are extremely expensive, and while they may provide estimates of the prevalence of FAS for small, local groups, they cannot provide accurate estimates for any particular region, much less for the United States as a whole.

A wide gulf remains between these numbers and other reported estimates. A 1997 study by the Seattle group claimed that the incidence of the syndrome in the general population was 91 cases per 10,000 births.[23] A project receiving commendation from the U.S. Public Health Service in *Public Health Reports* claimed — without attribution — that there are 5,000 FAS births annually in the United States and another 50,000 babies born with fetal alcohol effect.[24] Another study, published in the official journal of the National Institute on Alcohol Abuse and Alcoholism (NIAAA), reported the incredible statistic that "approximately 2.6 million infants are born annually following significant intrauterine exposure to alcohol."[25] Given the fewer than 4 million births in the United States in the year this study was published, this estimate would mean that *more than half* of all births in the United States are alcohol affected — a ludicrous claim. In the popular press likewise, reports of FAS have often exaggerated the extent to which the syndrome is increasing, with claims such as "rate of alcohol-injured newborns soars"[26] and "the percentage of babies born with health problems because their mothers drank during pregnancy increased six-fold from 1979 through 1993."[27] Indeed, the reported incidence of fetal alcohol syndrome has been increasing over time, from 1 per 10,000 births in 1979 to 6.7 per 10,000 in 1993, the most recent estimate for the nation as a whole.[28] However, because methods of identifying and tracking FAS have improved over this period, it is impossible to know whether the observed upward trend in incidence represents a true increase in the syndrome or merely reflects improved identification and reporting of cases.[29] Since FAS was still a new condition in 1979 and most doctors did not recognize it, this "sixfold increase" more likely represents an increase in the identification and reporting of cases than a true jump in incidence.

Perhaps because of these disparate estimates, even doctors have a hard time pinpointing how many children are affected by FAS. Most doctors could be fairly described as ignorant of the epidemiology of FAS; in my interviews, most admitted that they had no idea how many babies were affected. When they did volunteer an estimate of the number or rate of FAS births, the figures were often wildly inflated. "Should be a lot, should be hundreds of thousands," said one family practitioner. "Probably about 250 thousand to 400 thousand kids a year with somewhere between fetal alcohol effects and fully expressed FAS," asserted a pediatrician. "I imagine at [this hospital, with 2,500 to 3,000 deliveries a year] at any time [there are] at least one or two babies with FAS . . . in the nursery" [3F, male]. Given that most infants probably stay in the hospital for a few days at most, this last estimate implies a phenomenal rate of FAS births. Other doctors ventured estimates ranging from 2 percent of all births, to 4 to 5 percent, to 5 to 15 percent.

Despite their overestimates of the magnitude of FAS in the population, most of the physicians I interviewed believed that FAS was underdiagnosed, for a variety of reasons. First, the syndrome can be hard to see in a newborn. "I'm not sure how definite you can really call it in the nursery." It's not "going to be picked up necessarily by an obstetrician in the delivery room . . . [the] abnormalities are not apparent until someone reaches school age." Second, the "subtle changes" characteristic of FAS "can be difficult to diagnosis if you've never seen one before, you've never looked into it, or looked at a book about it, that's going to be a difficult thing to diagnosis from that standpoint." The diagnosis may be further complicated by doctors' reluctance to do the probing necessary to elicit a maternal history of alcoholism. "Physicians tend not to want to ask their patients about alcohol and drugs, especially in the private sector." Doctors acknowledged that "patients tend to underreport [alcohol use] and doctors don't probe more." After all, "who is going to admit it?" asked one pediatrician.

There are many reasons to suspect that even the most severe cases of FAS may go undiagnosed. Not only is the syndrome difficult to recognize from a clinical standpoint, it is also a hard diagnosis to make from an emotional one, as doctors reported. Moreover, children with FAS are unlikely to be living with their biological mothers. One study in Washington State, designed to increase referrals of children suspected of having FAS, found that only 20 percent were living with their biological mothers. More than half lived with foster or adoptive parents and the remainder lived with other biological relatives. The custody of the child can complicate making a diagnosis, since the maternal history of alcohol use during pregnancy — an essential diagnostic criterion — can be much harder to ascertain. Small, intensive studies like the one in Washington State have found that many children with FAS are not diagnosed as such; the referral program in Washington identified 39 children with FAS in a pool of 1,374 suspected cases. Only 1 of these 39 had previously been diagnosed with FAS.

Conversely, many children who do not have fetal alcohol syndrome may nevertheless be labeled with this diagnosis. Overdiagnosis is particularly likely in certain groups, such as American Indians. The reported incidence among American Indians is higher than among other groups, with rates as high as 98 per 10,000 among Southwest Plains Indians. Indeed, the reported incidence of FAS appears to vary greatly among different racial/ethnic groups. One study using data on birth defects collected by the CDC, for example, found the highest incidence among American Indians: 29.9 cases per 10,000 births. The rate among blacks (6.0 per 10,000) was six times higher than the rates among whites (0.9 per 10,000) and Hispanics (0.8 per 10,000). The reported incidence of FAS was lowest among Asian Americans (0.3 per 10,000).[30] Birth defects monitoring programs in Atlanta, Georgia, and in California also report sharp differentials in the rates of FAS among blacks and whites.[31] But because of the symptomatic imprecision of the syndrome and its predisposition to both overdiagnosis and underdiagnosis, it is difficult to tell what portion of the variation among population groups reflects actual differences in incidence and what portion is the result of diagnostic bias.

Thus, scientific evidence for three of the most common claims about fetal alcohol syndrome — that it affects a large number of births every year, that it affects some ethnic groups disproportionately, and that the incidence is increasing over time — is far from conclusive. The picture of FAS in the public imagination is colored by factors other than epidemiological "fact." "Perhaps the most straightforward way to establish a social problem's dimensions is to estimate the number of cases, incidents or people affected," argues Joel Best.[32] Like the cases of missing children examined by Best — indeed, like many other social problems — fetal alcohol syndrome is subject to overreporting and inflation. That tendency is exacerbated when the problem is one, like FAS, that lacks clear boundaries. One of the inflationary pressures on estimates of the incidence of FAS is the difficulty of defining the problem: what counts as alcohol-related birth defects, and therefore which cases do we need to count? Quantification is one technique by which social problems are reified. "The bigger the problem, the more attention it can be said to merit."[33] Gusfield provides a particularly compelling example of this phenomenon of inflation as it occurred in a single congressional debate about establishing the NIAAA in 1970, in which the magnitude of the problem ballooned from 5 million alcoholics to 18 million "problem drinkers" over the course of a single day.[34] What has happened with the diagnosis of so-called alcohol-related birth defects is very like what happened with the definition of alcoholism, in which attention shifted "from chronic alcoholism, with a pattern of addiction, loss of control, and frequency of intoxication, to a more diffuse and differentiated concept of 'problem drinking.'"[35] Because the relationship between alcohol exposure and outcome is so ambiguous, researchers can cast their nets ever more widely to capture the dimensions of the problem. FAS enjoys "the rhetorical advantages of broad definitions."[36] The scope of the problem depends on the definition of the syndrome or the range of congenital and developmental problems attributed to alcohol exposure in utero. "All statistics are the products of social processes. Someone has to decide what to count and how to go about counting."[37] Just as aggregate statistics are the product of social processes, so too are the individual cases. How doctors decide who has fetal alcohol syndrome and whether to label the child with a formal diagnosis is a process that may be subject to uncertainty, personal beliefs, and social bias.

Current Evidence on Women's Drinking Behavior

The belief that fetal alcohol syndrome is a widespread problem rests on the notion that drinking during pregnancy is common and frequent. Yet studies of drinking among women of childbearing age in the United States find that drinking is by no means universal and that drinking at levels that would be considered high risk for FAS is rare. Although we do not know with certainty the threshold above which alcohol damages the developing fetus, there is a consensus in the medical literature that consumption of one ounce of absolute alcohol (two typical drinks)[38] or more per day represents risk.[39] The CDC's Behavioral Risk Factor Surveillance System (BRFSS) found that, in 1995, almost half of women (51 percent) aged eighteen to forty-four did not drink at all, a very small proportion (3 percent) drank 7 to 14 drinks per week, and only 1 percent had more than 14 drinks per week. These findings are replicated in many studies of women's drinking behavior (see Table 6).[40]

Moreover, studies of women's drinking behavior during pregnancy confirm that only a small proportion drink heavily. In the 1995 BRFSS, 84 percent of pregnant women reported abstaining from alcohol and less than 1 percent reported having more than 14 drinks per week.[41] (About 3 percent reported binge drinking, which may also represent a risk of FAS.) According to data from 1990 birth certificates, fewer than 4 percent of mothers in that year reported drinking during pregnancy.[42] Unlike estimates of the incidence of FAS, studies of women's drinking behavior both during and outside pregnancy are remarkably consistent. Findings at the national level are echoed by smaller, local studies. Only about 2 percent of women enrolled in a California HMO reported regular drinking (defined as one or two drinks per week) during pregnancy.[43] Likewise, 4 percent of the women in a sample of black, inner-city women drank the equivalent of one or more ounces of absolute alcohol per day during pregnancy.[44]

Although birth certificate reports seem to be among the lowest self-reports of prenatal drinking behavior, *all* these studies report drinking behavior based on self-reports. The numbers may be underestimates if, as is likely, women are reluctant to admit to a socially condemned behavior.[45] In addition, pregnant or parturient women's retrospective reports of pre-pregnancy drinking behavior may be somewhat lower than reported drinking among all women of reproductive age, because women may modify their drinking behavior in anticipation of becoming pregnant. In the National Maternal and Infant Health Survey (NMIHS), for example, half the women reported that they wanted the pregnancy at that time, suggesting they may have been intending to become pregnant.

A high proportion of women who report drinking before pregnancy also report reducing or eliminating alcohol consumption while pregnant. In one study, 71 percent of pre-pregnancy drinkers reported abstinence at the eighteenth week of gestation.[46] In comparison, smoking during pregnancy is far more common: a national survey of young women found 8 percent who had one or more drinks a month while pregnant, compared with 40 percent who smoked.[47] Moreover, women who smoke are less likely to modify their behavior than are women who drink.[48] One study found that 53 percent of pre-pregnancy drinkers stopped drinking during pregnancy, compared with only 17 percent of smokers who stopped smoking.[49] Women who both smoke and drink are least likely to modify their behavior during pregnancy.[50] The strong correlation between smoking and failure to modify drinking behavior during pregnancy suggests that the observed symptoms of FAS may not be a function solely of in utero alcohol exposure but may depend on a constellation of related exposures.

Heavy drinking during early pregnancy increases the risk of spontaneous abortion,[51] so women whose fetuses are at greatest risk for alcohol-related birth defects may be less likely to experience a birth. In addition, women who drink heavily may elect to terminate their pregnancies through induced abortion — either because they are concerned about the risk their drinking poses to the fetus or because drinking behavior may be correlated with other characteristics that increase a woman's propensity to choose abortion (e.g., age, nonreligiosity), thus blurring the observed correlation between prenatal drinking and birth defects. If women who drink the most are less likely to experience a birth, then the observed effects of alcohol on the fetus will be attenuated.

One of the most puzzling aspects of the current literature on prenatal drinking behavior and FAS is the discordance between the observed incidence of the syndrome, which is concentrated in disadvantaged populations, and reports of drinking behavior. Drinking among women of childbearing age in the United States rises with income, education, and socioeconomic status; and white women are more likely to drink than minority women. A 1992 survey found that 48 percent of Hispanic women and 51 percent of black women said they did not drink at all, compared with 36 percent of white women. More-

Table 6. Estimates of Drinking among Women of Childbearing Age in the United States

Source	Population and Time Period	Category	Percent
Behavioral Risk Factor Surveillance System[a]	United States, 1995; women 18–44	Nondrinker	50.6
		<7 drinks/wk	45.7
		7–14 drinks/wk	3.0
		>14 drinks/wk	1.1
		5+ drinks on occasion	10.5
National Health Interview Survey[b]	United States, 1985; women 18–44	Any drinking	41.0
		>14 drinks/wk	3.0
Behavioral Risk Factor Surveillance System[c]	Alaska, 1991; women 18–44	Nondrinker	45.0
		≤30 drinks/mo.	38.0
		30+ drinks/mo	17.0
National Longitudinal Study of Women's Drinking[d]	United States, 1991; women 21 and older	Nondrinker	
		21–34	26.0
		35–49	35.0
		Heavy	
		21–34	4.0
		35–49	3.0
National Natality Survey[e]	United States, women giving birth in 1980	Any drinking	55.0
National Opinion Research Center Survey[f]	United States, 1981; women 21–49	Nondrinker	
		21–34	30.0
		35–49	28.0
		Light	
		21–34	41.0
		35–49	43.0
		Moderate	
		21–34	24.0
		35–49	20.0
		Heavy	
		21–34	6.0
		35–49	9.0
National Longitudinal Study of Women's Drinking[d]	United States, 1991; women 21 and older	Any drinking	58.0
		<3 drinks/wk	44.0
		4–13 drinks/wk	12.0
		14+ drinks/wk	3.0
Alcohol and Health Practices Survey[g]	United States, 1983; women 20 and older	Any drinking	50.0
		2+ drinks/day	4.0
National Maternal and Infant Health Survey[h]	United States, women pregnant in 1988	Any drinking	39.0
National Alcohol Survey 1992[i]	United States, 1992; women 18 and older	Nondrinker	
		White	36.0
		Black	51.0
		Hispanic	48.0

continued

Table 6. Continued

Source	Population and Time Period	Category	Percent
		Frequent	
		White	12.0
		Black	4.0
		Hispanic	3.0
		Heavy	
		White	3.0
		Black	5.0
		Hispanic	3.0
National Longitudinal Alcohol Epidemiologic Survey[j]	United States, 1992; women 18 and older	Alcohol abuse and dependence All races	
		18–29	9.8
		30–44	4.0
		Non-black	
		18–29	11.0
		30–44	3.9
		Black	
		18–29	3.3
		30–44	4.2

[a]Centers for Disease Control, "Alcohol consumption among pregnant and childbearing-aged women—United States, 1991 and 1995," *Morbidity and Mortality Weekly Report* 46, no. 16 (25 Apr. 1997): 346–50.

[b]Gerald D. Williams, Mary Dufour, and Darryl Bertolucci, "Drinking levels, knowledge, and associated characteristics, 1985 NHIS findings," *Public Health Reports* 101, no. 6 (Nov.–Dec. 1986): 593–98.

[c]Centers for Disease Control, "Linking multiple data sources in fetal alcohol syndrome surveillance—Alaska," *Morbidity and Mortality Weekly Report* 42, no. 16 (30 Apr. 1993): 312–14.

[d]Sharon C. Wilsnack, Richard W. Wilsnack, and Susanne Hiller-Sturmhöfel, "How women drink: epidemiology of women's drinking and problem drinking," *Alcohol Health and Research World* 18, no. 3 (1994): 173–81.

[e]Kate Prager, Henry Malin, Danielle Spiegler, Pearl Van Natta, and Paul J. Placek, "Smoking and drinking behavior before and during pregnancy of married mothers of live-born infants and stillborn infants," *Public Health Reports* 99, no. 2 (Mar.–Apr. 1984): 117–27.

[f]Richard W. Wilsnack, Sharon C. Wilsnack, and Albert D. Klassen, "Women's drinking and drinking problems: patterns from a 1981 national survey," *American Journal of Public Health* 74, no. 11 (1984): 1231–38.

[g]Charlotte A. Schoenborn and Bernice H. Cohen, "Trends in smoking, alcohol consumption and other health practices among U.S. adults, 1977 and 1983," *Advance Data* 118 (30 June 1986): 1–16.

[h]Analysis by author.

[i]Raul Caetano, "Drinking and alcohol-related problems among minority women," *Alcohol Health and Research World* 18, no. 3 (1994): 233–41.

[j]Bridget F. Grant, Thomas C. Harford, Deborah A. Dawson, Patricia Chou, Mary Dufour, and Roger Pickering, "Prevalence of DSM-IV alcohol abuse and dependence, United States, 1992," NIAAA's Epidemiologic Bulletin No. 35, *Alcohol Health and Research World* 18, no. 3 (1994): 243–48.

over, white women were three times as likely as black women to drink frequently (12 percent compared with 4 percent).[52] These estimates imply that we
should see higher rates of FAS among women who appear to be at greater risk
of drinking during pregnancy, but the observed incidence of FAS is concentrated among minorities and women of relative disadvantage (see Table 5).
How can we explain this disparity?

Several studies have found that the proportion of women who drink both
before and during pregnancy increases with income, education, and social
class[53] (see Table 6). For example, although the measured rate of FAS is some
six times greater among blacks than whites, studies indicate that white women
are more likely to drink during pregnancy.[54] In the 1980 National Natality Survey, the proportion of women drinking before pregnancy was higher among
whites (58 percent) than blacks (39 percent) and increased from 41 percent of
women who had not completed high school to 64 percent of college graduates.[55] Another study also found that among nonpregnant women, a higher
proportion of white women (57 percent) than black women (44 percent) reported drinking in the last month. The proportion drinking increased with education, from 47 percent of women with twelve or fewer years of education to
67 percent of college graduates. Fifty-one percent of married women drank,
compared with 62 percent of unmarried women. Both pregnant and non-
pregnant smokers were more likely to drink than nonsmokers.[56] Looking at
women's drinking behaviors during pregnancy as a means to understand the
distribution of risk of FAS is not without its flaws. Drinking is an underreported
behavior and is likely to be particularly so during pregnancy, due to social de-
sirability bias.[57] Moreover, drinking during pregnancy tends to be correlated
with other behaviors and characteristics that put the baby at risk of an adverse
outcome. Drinking during pregnancy is certainly not a perfect measure of FAS
or even of the risk of FAS, but estimating the extent of this behavior can help
to fill in our understanding of the distribution of risk.

Drinking Behavior in the National Maternal and Infant Health Survey

The 1988 NMIHS was a nationally representative survey of women between
the ages of fifteen and forty-nine who experienced a pregnancy in 1988. It included information on the amount of alcohol consumption before and during
pregnancy, why the woman reduced drinking during pregnancy, and whether

she received prenatal care and medical advice to stop or cut back her drinking. (For a full description of the sample and methodology, see the appendix.)

Not all women of reproductive age drink; I assume that women who do not drink at all before pregnancy are consequently not at risk of drinking during pregnancy. In order to understand patterns of drinking behavior during pregnancy, we need first to look at drinking among women who may become pregnant. As shown in Table 7, a substantial minority — 39 percent — of women reported drinking in the twelve months before the pregnancy.[58] However, the proportion of women who reported drinking varies by demographic characteristics. Almost half of white women reported drinking, compared with less than a third of black women. The proportion of women drinking rises sharply with educational attainment, from 28 percent among women with less than twelve years of schooling to 36 percent among high school graduates, 47 percent among women who have attended college, and 63 percent among women with graduate or professional schooling. Drinking also rises with age, from 24 percent among women less than twenty years old to 43 percent among women thirty-five and older. The largest proportion of drinkers is among women thirty to thirty-four years old; almost half of these women reported drinking before pregnancy. Fewer unmarried women reported drinking than married women. As with education, the proportion of women who drink rises with income. About a third of women with family incomes of less than $20,000 reported drinking, compared with 41 percent of women with incomes between $20,000 and $35,000 and 56 percent of women with family incomes above $35,000. Pregnancy intentions do not appear to influence the likelihood that a woman drinks before pregnancy. In other words, women who are trying to become pregnant or want to become pregnant appear no more or less likely to drink than other women of childbearing age.[59] These findings echo those of other studies of women's drinking behavior: women of relative social advantage are more likely to drink.

Yet these differences in drinking behavior are much less marked when considering reduction of alcohol consumption during pregnancy. Among the women who drank before pregnancy, almost all (89 percent) reported reducing drinking during pregnancy. Moreover, there is little variation by maternal characteristics in the proportion of drinkers who reduced alcohol consumption. In all groups, the majority of women reported reducing drinking during pregnancy (see Table 7).

Table 7. Drinking before Pregnancy and Reduced Drinking during Pregnancy, 1988 National Maternal and Infant Health Survey

	Percentage of Women Reporting	
	Drinking before Pregnancy (N = 9,953)	Reduced Drinking during Pregnancy (N = 3,881)
All	38.8	88.8
Mother's race		
White	49.2	91.5
Black	29.1	84.0
Other	26.2	91.5
Years of education		
<12	28.4	81.8
12	36.2	88.3
13–16	47.2	91.6
≥17	63.3	94.6
Mother's age at birth		
<20	24.1	85.8
20–29	40.2	89.7
30–34	47.8	89.6
≥35	42.7	86.6
Marital status		
Married	43.1	90.7
Unmarried	32.5	85.3
Parity		
1	38.5	92.1
2	39.1	90.4
≥3	39.0	79.5
Household income		
<$10,000	30.3	83.5
$10,000–$19,999	33.2	89.3
$20,000–34,999	40.5	90.2
≥35,000	55.8	91.9
Pregnancy intentions		
Wanted	40.9	91.0
Mistimed	37.4	88.6
Not wanted	33.6	75.6

Many clinicians speak of pregnancy as a window of opportunity, a period in which women are especially motivated to change such health behaviors as drinking, smoking, and diet. And women in the NMIHS overwhelmingly cited concern for the health of the baby as the reason they reduced drinking (see Table 8). By far the most important reason is concern for the health of the

Table 8. Reasons for Reduced Drinking during Pregnancy,
1988 National Maternal and Infant Health Survey

Reason	Percentage of Women Reporting[a]	
Concerned about baby	85.4	
Aversion		
Lost desire to drink	20.7	
Alcohol made her sick	8.8	} 38.0
Alcohol tasted/smelled bad	7.6	
Urged by doctor	18.9	
Urged by family	8.0	

[a]Women may report more than one reason, so percentages do not total 100.

baby — 85 percent of the women cited this as a reason for their behavioral change. The second most important reason, cited by 38 percent of the women, was that they developed an aversion to alcohol; half of these women reported that they simply lost the desire to drink. Fewer than 20 percent said they changed their alcohol consumption at the urging of a doctor. Even smaller numbers of women reported that alcohol made them sick, that it tasted or smelled bad, or that they were urged by their families to give up drinking.

To assess the risk of FAS, we need to know not only whether a woman drank at all during pregnancy but how much she drank. Ideally, we would like to know both how much she drank and the timing of alcohol consumption during gestation, but these pieces of information are generally unavailable. The NMIHS provides information on the number of drinks a woman had per week. Although we do not know definitively how much drinking contributes to the expression of FAS, risk levels are most often conceptualized in terms of the number of drinks per day. Having 2 or more drinks per day (14 or more drinks per week) is considered high risk; fewer than 3 drinks per week is classified as low risk. Thus, medium risk is characterized here as 3 to 13 drinks per week. One shortcoming of the NMIHS is that it does not provide any information on the amount of alcohol consumed per drinking episode. Thus, a woman who reported having 5 drinks per week may have had 1 drink every night or may have had a drinking binge one night a week. These patterns represent potentially very different risks for fetal alcohol syndrome. Moreover, the NMIHS, like almost all studies of drinking behavior, relies on self-reports to measure alcohol consumption levels.

Only a very small minority of women reported consuming alcohol at high

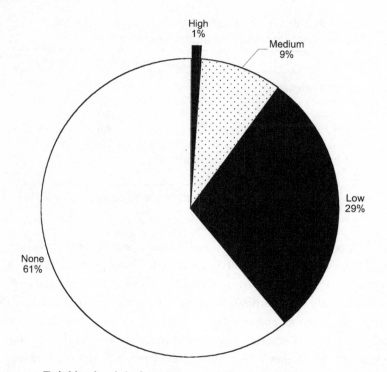

FIGURE 5a. Drinking levels before pregnancy, all women: high, 14 or more drinks per week; medium, 3 to 13; low, fewer than 3 drinks per week.

risk levels, as I have defined them. As seen in Figure 5a, among the entire sample of women, only about 1 percent reported consuming 14 or more drinks per week before pregnancy. Even among the subset of women who reported any drinking before pregnancy, only 4 percent reported drinking at risky levels. These proportions give some indication of the potential prevalence of high-risk drinking during pregnancy. In fact, women who drink at high risk levels constitute an even smaller proportion during pregnancy. Only 1 percent of pre-pregnancy drinkers reported drinking this amount during pregnancy (see Figure 5c), and about 0.5 percent of women in the overall sample reported drinking 14 or more drinks per week during pregnancy (Figure 5b). Moreover, 83 percent of all women reported no alcohol consumption while pregnant (Figure 5b), and more than half (56 percent) of the women who reported pre-pregnancy drinking said they did not drink at all during pregnancy (Figure 5c). These numbers suggest not only that drinking at high risk levels during preg-

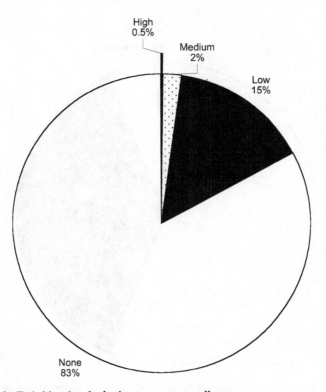

High
0.5%

Medium
2%

Low
15%

None
83%

FIGURE 5b. Drinking levels during pregnancy, all women

nancy is rare but that most women modify even medium- and low-risk drinking behaviors while pregnant.

However small their number, women who drink at high risk levels before
pregnancy are potentially most at risk for delivering babies with fetal alcohol
syndrome if they continue drinking at the same levels during pregnancy. Figure 5d provides evidence of the extent to which these women changed their
drinking behavior when pregnant. The figure shows the proportion of pre-
pregnancy high-risk drinkers who continued to drink at high risk levels, as well
as the proportions who reduced or eliminated drinking.[60] Overall, only one-
fourth of the women continued to drink at high risk levels and one-fifth eliminated alcohol consumption entirely while pregnant (Figure 5d).

Moreover, pregnancy intentions are correlated with the level of drinking
during pregnancy. Sixty percent of women who wanted the pregnancy at the

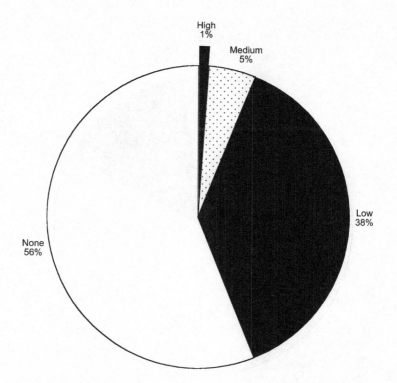

FIGURE 5c. Drinking levels during pregnancy, women who drank before pregnancy

time of conception stopped drinking during pregnancy, compared with 54 percent of women who wanted to delay pregnancy and 45 percent of women who had no intention of ever becoming pregnant. A higher proportion of these women who did not intend pregnancy continued drinking at moderate to high risk levels during pregnancy: 15 percent reported consuming 3 to 21 drinks per week, compared with 4 percent among women who wanted to become pregnant at the time they conceived. Among high-risk drinkers, 80 percent reported that they did not intend to get pregnant when they conceived. Unintended pregnancy thus appears to raise the risk of drinking at high risk levels prenatally and may therefore increase the risk of FAS.

The evidence presented here suggests some new ways of thinking about the risks of drinking during pregnancy. (Keep in mind that the NMIHS data were collected in 1988, before the alcoholic beverage label law went into effect.) First, we see that women who are white, women who have more educa-

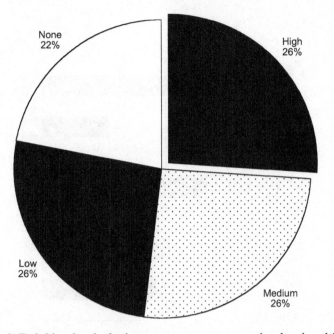

FIGURE 5d. Drinking levels during pregnancy, women who drank at high risk levels before pregnancy

tion, and women with higher family incomes are more likely than other women to drink *before* pregnancy. Second, we see that among women who drink before pregnancy, almost all stop or reduce drinking during pregnancy. Drinking during pregnancy is *not* a widely prevalent behavior in this representative sample of American women. Third, we learn that a very small proportion of women drink at what could be considered high or medium risk levels for FAS. Among the tiny minority of women who said that they drank during pregnancy, the vast majority reported having fewer than three drinks per week.

To reiterate, most women do not drink while they are pregnant. Among those who do, most do not drink at levels that would be considered to pose an appreciable level of risk. There is yet another way of looking at the data — through statistical methods known as multivariate logistic regression. This method enables us to consider the likelihood — or the odds — that a woman with particular characteristics will engage in a behavior such as drinking, relative to a reference or comparison group. In Figures 6 and 7, the bars represent the odds that a woman drinks. For each category (race, education, age at

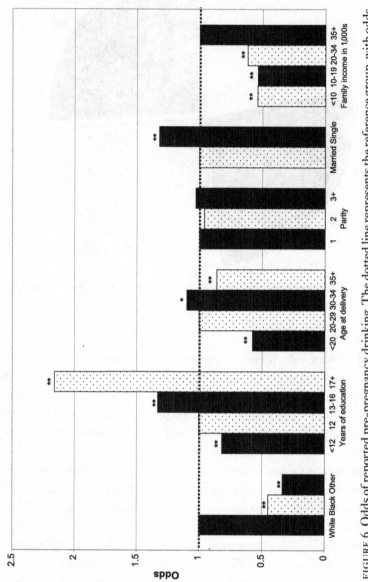

FIGURE 6. Odds of reported pre-pregnancy drinking. The dotted line represents the reference group, with odds of 1.0. * $p < 0.05$; ** $p < 0.01$.

delivery, parity, marital status, and family income), the comparison subgroup, or reference group, is set at an odds of one (shown by the dotted horizontal line). Thus, if other subgroups in that same category have odds greater than one (bar rising above the reference line), they are more likely to drink than women in the reference group. Conversely, if the odds are less than one (bar falling below the line), women in the subgroup are less likely to drink than women in the reference group. The asterisks above the bars indicate whether the difference between each subgroup and the comparison group in that category is statistically significant. (For detailed data tables, see the appendix.)

Figure 6 presents the odds of reporting drinking before pregnancy among all women in the sample. Black women are less than half as likely as white women to drink before pregnancy. Other minorities (primarily Asian Americans and American Indians) are only two-thirds as likely to drink as white women. The odds of drinking increase with education and are more than twice as high among women with a graduate degree as among women who did not complete high school. And women who have not completed high school are about 20 percent less likely to drink than high school graduates. The odds of drinking are highest among women aged thirty to thirty-four and lowest among teenagers. Differences in pre-pregnancy drinking levels by parity (number of children already born) are negligible. However, unmarried women are 40 percent more likely to drink than married women. The odds of drinking increase with income.

Figure 7 shows the odds of drinking at high risk levels (14 or more drinks per week) during pregnancy, for women who drank before pregnancy. Because, in this chart, the bars that rise above the dotted line represent the relative likelihood that women in different subgroups will drink at levels that may place their children at greater risk of FAS, Figure 7 provides crucial information. First we should note some sizable reversals in the odds associated with various characteristics. For example, black women are less likely than white women to drink before or during pregnancy or to drink at high risk levels before pregnancy. However, their odds of drinking at high risk levels *during* pregnancy are twice those of white women. Similarly, less-educated women have higher odds of high-risk drinking during pregnancy, despite their lower odds of drinking overall, and women who have attended college are only half as likely as high school graduates to drink at high risk levels, despite their higher odds of drinking before pregnancy. Although women with more education are more likely to drink before pregnancy, they are also more likely to reduce al-

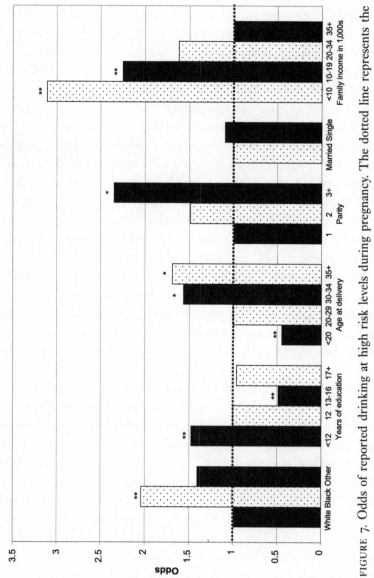

FIGURE 7. Odds of reported drinking at high risk levels during pregnancy. The dotted line represents the reference group, with odds of 1.0. * $p < 0.05$; ** $p < 0.01$.

cohol consumption while pregnant. The odds of drinking while pregnant increase with age and parity in a pattern similar to that observed for drinking before pregnancy. In fact, women aged thirty-five and older are almost 70 percent more likely to drink at high risk levels than women aged twenty to twenty-nine, and women pregnant with their third or subsequent child are more than twice as likely to drink at high risk levels as women pregnant with their first child. The odds of high-risk prenatal drinking are higher among women with lower income levels than among women with family incomes of $35,000 or more; however, the odds fall as income rises.

These two bar charts, along with the data presented earlier in the chapter, provide fresh insight into the disparity between reports of drinking behavior — higher among white, more highly educated, married, nonpoor women — and the incidence of FAS — higher among poor, minority, and less educated women. Women who initially exhibit the greatest tendency to drink are also most likely to reduce alcohol consumption during pregnancy. Thus, the reversals in odds in drinking behavior before and during pregnancy can be attributed to women's differential exiting from the "risk set"; white women, more educated women, and women with higher incomes, although more likely to drink initially, are also more likely to reduce drinking while pregnant, leaving the risk set. Because women of relative disadvantage are less likely to modify their drinking during pregnancy, they end up comprising the group most at risk. Black women, women with less than twelve years of schooling, older women, and higher parity women thus constitute the group most likely to drink at high risk levels during pregnancy; therefore, their babies are at greatest risk of FAS. These maternal characteristics do indeed reflect the observed incidence of FAS in the population. It is a paradox that women who appear to be at the lowest risk initially have the highest odds of high-risk drinking during pregnancy.

Fewer than half the women interviewed in the NMIHS reported drinking before pregnancy, and almost all reported reducing alcohol consumption while pregnant. However, a small minority of women did not change their high-risk drinking behaviors during pregnancy. The correlates of high-risk drinking during pregnancy — black race, limited education, advanced age, high parity, and poverty — are factors that themselves predispose women to adverse birth outcomes. Moreover, despite their higher levels of risk, these women are least likely to receive medical advice to reduce drinking during pregnancy.[61]

Rhetoricizing Risk: FAS in the Public Imagination

Fetal alcohol syndrome is a complex phenomenon, little understood from a demographic perspective. This chapter has focused on discerning the women at greatest risk and has demonstrated that the determinants of high-risk drinking during pregnancy are not readily apparent from observations of drinking behavior among all women. Two of the central tenets of the social construction of FAS are that it is a "big" problem and that all pregnant women are at risk. Yet the picture drawn by the analysis above suggests that the conventional wisdom may be wrong on both counts. Drinking during pregnancy — and thus the risk of FAS — is not only a rare occurrence but a highly concentrated one as well.

Despite reports suggesting that alcohol-related birth damage is a social problem both deep and wide, fetal alcohol syndrome is concentrated in certain populations and among children of women with a nexus of characteristics beyond drinking alone, such as smoking, poverty, malnutrition, or environmental exposure. Prospective studies have found that only 5 to 6 percent of the infants of alcoholic women are born with fetal alcohol syndrome,[62] and that FAS is highly correlated with advanced maternal age, high parity, poverty, smoking, and drinking patterns such as type of alcohol, timing of fetal exposure to alcohol, and peak blood alcohol concentration levels reached per drinking episode.[63] Moreover, there is much evidence to suggest that the causal link between alcohol and teratogenesis is a complex one. Recently, scientists have argued that fetal alcohol syndrome may not occur through alcohol exposure alone; rather, a range of "permissive" or "provocative" cofactors must be present to cause the defects associated with FAS.[64] These include such things as undernutrition, exposure to environmental pollutants, extreme stress, and tobacco use. Other studies have found no association between congenital malformations and alcohol consumption, even among the heaviest drinkers.[65] Early reports of FAS were retrospective and "described highly selected populations."[66] These early reports of the sequelae of drinking during pregnancy may thus have failed to consider fully the presence of other contributing risk factors such as smoking, inadequate nutrition, high levels of stress, advanced age, or high birth order, consequently overestimating the contribution of alcohol to poor birth outcomes.

The attention to alcohol-related "birth damage"[67] — disproportionate to its actual incidence in the population — both reveals our society's quest to control reproductive outcomes and illustrates the tendency to see disease as the consequence of individual behavior. The varying estimates of the incidence of FAS reflect the difficulty in detecting the condition and the arbitrary nature of the diagnosis. Since there is no biological marker, no lab value that can confirm or refute a diagnosis of FAS, the syndrome is ultimately, inevitably, socially constructed. It exists in the eye of the beholder and is particularly susceptible to detection bias. Medical professionals may be predisposed to see FAS in the children of women known to drink and may ascribe the behavior of prenatal drinking to certain groups, which are then labeled at "high risk" of FAS,[68] as American Indians have been.[69]

In some cases, too, there may be a deliberate if well-intentioned misreporting of the numbers. Two articles, one from the popular press, the other from the public health literature, illustrate the claims typically made about the incidence of FAS and the ways in which "large" numbers are used to "accentuate significance and urgency."[70] A front-page *Philadelphia Inquirer* article on truancy among teenagers of the Omaha Indian tribe claimed that "a third or more of the Omaha's children suffer from FAS."[71] The reporter described FAS as "physical and *emotional* retardation caused by drinking during pregnancy" (emphasis added). Here the reporter has vastly expanded the domain of those who suffer from FAS and further demonstrated how the condition may be used to make sense of painful social realities, in this case the alienation felt by Omaha youth. Often, FAS serves as a way to explain extreme social disintegration, as it does here.

Similarly, an article on the pediatrician's role in preventing alcohol abuse, published in *Alcohol Health and Research World*,[72] the official journal of the NIAAA, claimed that "approximately 2.6 million infants are born annually following significant intrauterine exposure to alcohol."[73] In the first instance, an exaggerated estimate of the prevalence of FAS is used to explain social deviance. In the second, a high incidence of FAS is used to create a sense of urgency, to justify a proposed public information campaign, and to dramatize the need for pediatricians to heed FAS.

Public health advocates may also overstate the incidence of FAS because they believe that existing measures of FAS underestimate its true incidence. Indeed, the term *fetal alcohol effect* was created as an official label for alcohol-related negative birth outcomes for which a diagnosis of FAS was not legitimate, because one or more of the necessary conditions for diagnosis were absent. The criteria for a diagnosis of FAE, or partial FAS, or "alcohol-related birth defects" are ill-defined; these labels may be used quite loosely for any child who has problems and whose mother drank during pregnancy.[74] Thus, the FAE label greatly expands the universe of children affected by prenatal drinking. The National Organization on Fetal Alcohol Syndrome (NOFAS) claims, for example, that "as many as 12,000 infants are born each year with FAS and three times as many have ARND or ARBD. FAS, ARND and ARBD affect more newborns every year than Down syndrome, cystic fibrosis, spina bifida and Sudden Infant Death Syndrome combined."[75] Gusfield points out the way in which "such a statement of magnitudes, although it creates a factual world of order and certitude, is in another sense a statement of rhetoric — a way of saying: Look, this is an important problem. It deserves attention and priority."[76]

Fetal alcohol syndrome is commonly referred to as the "leading preventable cause of birth defects." The meanings implicit in this claim warrant careful examination. Most congenital malformations are not caused by known teratogens — in fact, we know the cause of only about 25 percent of birth defects[77] — but rather are genetic accidents or cases of fetal development gone awry. The exaggerated status attributed to FAS by naming it as the leading preventable cause of birth defects, like inflated estimates of the incidence of FAS, serves to create an aura of urgency and gravity around the problem. In fact, the medical literature on FAS often falls short of objective, dispassionate writing and relies on moralistic rhetoric. Witness an article in the journal *Minnesota Medicine* that claims to document the "many shocking characteristics" of children with FAS and to "reveal the devastating problems *victims* of FAS and FAE inherit" (emphasis added).[78] This language sounds more like a carnival sideshow pitch than a clinical case presentation.

The problem of FAS is inflated not only through exaggerated estimates of its incidence but also, more bluntly, through repeated assertions of the gravity of the problem. The Partnership to Prevent Fetal Alcohol Syndrome, which is sponsored by the federal Substance Abuse and Mental Health Services Administration (SAMHSA), claims that "FAS and ARBD are . . . among the most critical health issues facing communities today."[79] A 1997 book on FAS began with a similar claim. Washington State Governor Mike Lowry wrote in the foreword, "I believe that research on Fetal Alcohol Syndrome and Fetal Alcohol Effects clearly is one of the most important issues in all the world today."[80]

Moreover, the oft-repeated contention that FAS is *preventable* reflects a social desire to control reproductive outcomes for the better. Our society expresses a deep uneasiness about less than optimal birth outcomes and likewise exhibits a desire to eliminate such outcomes — in fact, to eliminate even the *risk* of such outcomes.[81] The battery of tests to detect abnormalities is rapidly expanding, even as the Human Genome Project attempts to find and label all genetic sources of biological (and even, in such cases as the "alcoholism gene" or the Y-linked "violence gene," social) deviance. FAS is yet another manifestation of our cultural fixation on explanation and prevention, our need to bring randomness within the purview of our control. By labeling FAS and ruling it preventable, we reassure ourselves that we can eliminate all variance in birth outcomes and produce only "perfect" babies. One of the family practitioners I interviewed spoke very movingly about our society's tendency to devalue completely those birth outcomes that are less than optimal. "I think our culture . . . really routinely feels that if people are damaged, then they are tragic. So that children like my daughter who has Down's syndrome are perceived as refuse by our culture and . . . killing them when they're in utero is the desirable outcome . . . Our culture is addicted to accomplishment." As advances in contraception have afforded women ever greater control over the number and timing of children they will bear, and as prenatal diagnostic technology has expanded, the impulse to control the quality of births has intensified.

Ironically, public policy to prevent FAS is probably singularly ineffective in actually reducing the incidence of alcohol-affected births. To date, efforts to prevent FAS in the United States, such as the federally mandated alcoholic beverage label law, have been cast widely, attempting to reach all pregnant women regardless of actual risk. Nineteen states and many localities (including California, Arizona, Georgia, Oregon, Illinois, Tennessee, Minnesota, Washington, West Virginia, New Jersey, New York City, Philadelphia, Racine, Wisconsin, and Washington, D.C.) also require bars and restaurants to post warnings about drinking during pregnancy.[82] In New Jersey these posters read, "A pregnant woman never drinks alone."

Such policies not only target the vast majority of women unnecessarily, they may actually neglect the populations most adversely affected by prenatal drinking. Study after study has found that the alcoholic beverage label law has only a small impact on drinking during pregnancy.[83] In fact, most studies have found that changes in drinking behavior typically occur only among light drinkers, defined as those who have less than one drink per day.[84] These findings are consistent with other studies reporting that high-risk drinkers are less responsive to media campaigns.[85] Not only the amount that a woman drinks but also the kind of alcohol she consumes may affect her awareness of the label. Women who drink beer or wine coolers are more likely to have seen the label than women who drink wine or liquor.[86] In American Indian communities, which bear the highest burden of alcohol-related morbidity, warning labels and signs cannot be effective on officially dry reservations or where home-brewed alcohol is commonly consumed.

Such broad-based public prevention efforts fail on two levels. On one, they target many women unnecessarily, potentially increasing the stress of pregnancy. The notion that excessive warnings may alarm women and be potentially counterproductive is commonly expressed in the European literature on FAS yet almost entirely absent from the American literature — although many doctors whom I interviewed were cognizant of this consequence of public education efforts. The American policy response, which focuses on nothing short of total abstention, arguably has roots in this country's earlier temperance movements.

On a second level, public warnings do nothing to address the actual medical and social needs of the populations most adversely affected by prenatal drinking. Moreover, these public warnings serve as constant reminders to women that they bear responsibility for the well-being of the next generation. As such, the warnings are a diffuse kind of social control, a message transmitted to *all* women, pregnant or not, that they ought to live their lives within the compass of a nurturing maternity.

For all its ineffectualness in addressing the problems behind FAS, our public policy is a highly effective expression of social control. The spectrum of social anxieties and concerns captured in the ways we think about FAS and risk during pregnancy has blunted the ability of policy to target the women most at risk of adverse birth outcomes. The analysis presented in this chapter has helped to demonstrate the ways in which the social construction of FAS — and subsequent public policy — is incongruent with empirical reality. One consequence of this incongruity is that all women are made to suffer unnecessarily a burden of guilt and responsibility. Another is that the women and children whose lives are most adversely affected by alcohol are essentially ignored. And perhaps the gravest consequence of all is that public attention is directed away from the social inequities that are at the root of poor birth outcomes and towards a rhetoric of individual responsibility and blame that privileges private morality over public justice.

CHAPTER 6

MEDICAL-MORAL AUTHORITY AND THE REDEFINITION OF RISK IN THE TWENTIETH CENTURY

Alcohol and pregnancy.
No safe time. No safe amount. No safe alcohol.
Period.

—NOFAS slogan

Despite the scant initial evidence, despite the infrequency of heavy drinking during pregnancy, and despite the continuing skepticism of some doctors about the diagnosis, fetal alcohol syndrome and its associated conditions are today well established and influential in the sphere of medicine and in the world of public policy — and in the private experience of pregnancy of millions of women in the United States every year. How rapidly and how deeply FAS and the risk attributed to drinking during pregnancy took root in American medical and popular imagination leaves us with several questions. What prompted the discovery of FAS in 1973? And in the decades following, how did this initially obscure diagnosis of a set of severe birth defects among the children of chronic alcoholic women who drank to excess during pregnancy transmogrify into a pervasive risk assumed to threaten *every* pregnancy in the United States and to cause a wide spectrum of purported adverse consequences, from frank mental retardation to learning disorders, from major organ malformation to minor motor disabilities? In this chapter I consider why doctors suddenly shifted course in their beliefs about alcohol and pregnancy, returning in the early 1970s to the question of alcohol and offspring that they had attempted to answer a century earlier, before abandoning it altogether. And I attempt to explain the public response to this new medical knowledge, a response that in many ways seems both disproportionate to the evidence of risk and maladapted to the actual problem. How do social problems, such as the fear about drinking during pregnancy, become miscalibrated in this way?

The emergence of the diagnosis of fetal alcohol syndrome demonstrates some of the ways in which disease may be used to exert social control, at once expressing and reinforcing social ideologies. Disease provides a powerful lens through which to view society and social conflict. "Because diseases often evoke and reflect collective responses, their study can provide an understanding of the values and attitudes of the society in which they occur."[1] Indeed, the diagnosis of FAS both "evokes and reflects" a particular era of American social history. The discourse of rediscovery, so crucial to the early construction of the diagnosis, also illustrates the cycles of attention and inattention that infuse the history of medicine.[2] This latest cycle of attention to drinking and birth outcome emerged at a time when awareness of teratogenic substances in the environment was increasing, when the fetus was gaining medical and popular visibility and status, and when the desire to control reproductive outcomes was becoming stronger. Following the sexual revolution of the 1960s, pregnancy and reproduction became highly politicized in the 1970s. Not coincidentally, gender relations were undergoing a significant transformation at the same time. Finally, American society was experiencing a general cultural retrenchment and growing conservatism that affected how alcohol and social problems were viewed. These broader social forces, as well as trends within medicine, contributed to the sudden visibility of fetal alcohol syndrome in 1973 and the escalation of concern about drinking during pregnancy in the years since.

A Poisoned Environment

In the world at large, and particularly in the United States, a growing awareness of environmental risks was building throughout the 1960s and 1970s. The publication of Rachel Carson's *Silent Spring* in 1962 galvanized public attention to environmental toxins. People felt a heightened sense of threat to personal health from toxic and teratogenic agents in the environment. The late 1960s and early 1970s were a period of intense alarm about the environment generally and human exposure to harmful synthetic substances such as DDT and PCBs particularly. The Toxic Substances Control Act was first introduced into Congress in 1969 (and passed in 1976), and in 1970 Congress enacted major amendments to the Clean Air Act. In fact, environmental concerns were often constructed as or transmuted into threats to human health, thus personalizing risk.[3]

Moreover, the public sense of environmental endangerment became linked to a growing awareness of the vulnerability of the fetus, as a result of two events in the early 1960s. From 1959 to 1962, thousands of women around the world, primarily in Europe, gave birth to babies with severe deformities after taking the prescription sleeping drug thalidomide during pregnancy. The babies were born with no or only vestigial limbs — a condition known as phocomelia — as well as crippling internal organ malformations. In the United States the thalidomide disaster was largely averted through the vigilance of the FDA; some children whose mothers had participated in research protocols were born with the phocomelia caused by thalidomide, but since the drug was never approved for use, the disaster was largely contained.[4] However, in 1961, unforgettable images of the thalidomide children appeared on television and throughout the print media in the United States and worldwide, helping to convince both doctors and women that the fetus was at risk in ways they had not previously imagined.[5] Ashley Montagu's popular 1965 book *Life before Birth* epitomizes the effect of the thalidomide disaster on lay and medical thinking. "For those of us in the world who were spared, the story of thalidomide is a warning bell, clanging out an urgent warning: it can be dangerous to gobble down *any* pills indiscriminately and unthinkingly. We depend on our Food and Drug Administration and our doctors to protect us from harm from drugs, but this is not enough. Each of us must take personal responsibility, too, for ourselves and our children" (emphasis in original).[6] Montagu's theme of "personal responsibility" is one that has become increasingly dominant in medicine over the last three decades and is an important element in the social backdrop to the discovery of fetal alcohol syndrome.

Shortly after the horror of the thalidomide babies, in 1964–65 the United States experienced a small rash of babies born with profound disabilities after their mothers' exposure to rubella, or German measles, during the first sixteen weeks of pregnancy. Both events helped to change the way the lay public and doctors alike thought about pregnancy and the vulnerability of the fetus. The placenta could no longer be seen as an impermeable barrier. Just as the environment became an unsafe place, full of hazardous toxins against which the public was powerless to defend itself, the womb came to seem increasingly threatened by external forces and by substances the pregnant woman was exposed to or even ingested — pills she gobbled down "indiscriminately and unthinkingly," in Montagu's memorable phrasing. (It is hard to miss the irony of Montagu's warning about thalidomide and other "pills" in contrast with his reassurance of the absolute safety of alcohol.) Pregnancy now seemed not only perilous but imperiled. The frontispiece for Montagu's *Life before Birth*, for example, proclaimed, "From the moment of conception until delivery nine months later, the human being is more susceptible to his environment than he will ever be in his life again." This sentiment echoes strongly the concerns of the late nineteenth century. Just as then, both doctors and the public evinced growing worry about the quality of births and invested the prenatal period with extraordinary significance.

Moreover, just as the environmental movement raised public consciousness and awareness of risk and contributed to the sense that nothing in the world was safe any longer, the "discovery" in 1962 of child abuse as a medical problem, initially labeled "battered child syndrome,"[7] contributed powerfully to the dissolution of old ideas about parents as protectors of their children. Social, medical, and political attention to child endangerment intensified rapidly in the years before and just after the discovery of fetal alcohol syndrome. Once physicians and the public could "see" a world in which mothers or fathers might hurt their living children, it became easier to think of a pregnant woman harming or abusing her unborn child. Indeed, the specter of child abuse is often evoked in the popular and the medical literature about FAS, with disturbing images of children "wounded"[8] or "bruised before birth"[9] or even explicit characterizations of FAS as "child abuse in the unborn fetus."[10] This conceptualization of FAS as child abuse became literal in 1990, when Wyoming became the first state to charge a pregnant woman who was drunk with felony child abuse.[11] The view of drinking during pregnancy as a form of child abuse made possible the assault and attempted homicide charges against Deborah Zimmerman (see Chapter 1).

The Newly Visible Fetus

Significant changes in medicine and medical knowledge also helped set the stage for the discovery of FAS. Most notably, doctors began to see pregnancy in a new light, not only as a result of greater awareness of the threats to pregnancy. The decades just before 1973 witnessed increasing medical attention to quantifying and guaranteeing the quality of all births, damaged or not. The introduction of Virginia Apgar's eponymous score in 1952 gave doctors a formalized and universal way to assess instantly the viability and well-being of every newborn in the delivery room. Attention to birth defects surveillance intensified in the 1960s and early 1970s following both the thalidomide disaster in Europe and the birth of babies affected by the 1964–65 rubella epidemic in the United States. In fact, it was in 1973 that the Centers for Disease Control established the national Birth Defects Monitoring Program to measure and keep track of anomalous births throughout the country.[12]

Indeed, in the two decades prior to 1973 doctors were engaged in an intensive effort to identify, classify, and categorize patterns of birth defects. For physicians, anomalous births represented a disorder that was at once physical *and* analytical. David W. Smith, one of the discoverers of FAS, published the first comprehensive atlas of birth defects in 1970 and followed it with a second edition in 1976. Of the 225 named defects listed in the second edition, more than half were identified between 1950 and 1974, and 75 were named in the 1960s alone. Thus, Smith and his fellow dysmorphologists were engaged in a quest to impose the order of diagnosis on a wide range of congenital and genetic patterns of malformation and dysfunction. In discovering and naming fetal alcohol syndrome, Smith was merely continuing his life's work of ascribing etiology to disorder.

Not only were physicians in this period more keen to understand and classify anomalous births, but new technologies were producing profound changes in the medical treatment of pregnancy and birth. Most revolutionary were ultrasound, first used in pregnancy in 1958, and amniocentesis, first used to diagnose fetal chromosomal anomalies in 1967. Both these technologies afforded doctors a view of — and into — pregnancy and fetal development that they had never had before. Moreover, development of new technologies like amniocentesis gave doctors a new sense of control over undesirable birth outcomes. Amniocentesis enabled doctors to predict which births would be affected by gross anomalies — which in turn allowed women to choose to terminate those pregnancies. Indeed, the advent of prenatal genetic testing precipitated a profound shift in how doctors, women, and society at large viewed babies with birth defects. What had previously been regrettable now became "preventable." Babies with birth defects were now seen not merely as tragedies but as tragedies that could be prevented — or at least averted — through medical intervention.

In some respects, babies born with birth defects became even less socially accepted, their mothers more stigmatized. Over the last three decades, pressure has been building on women to accept prenatal diagnostic technology; this pressure is accompanied by implicit blame for women who refuse testing and whose babies are born with defects.[13] Women who fail to prevent such damaged births are seen as culpable. Thus, prenatal diagnostic technologies occasioned a subtle shift in notions of responsibility towards greater individual responsibility and moral culpability of the woman who bore a child with defects. When birth defects are viewed not as random accidents of fate but as events preventable by human agency, responsibility for defective babies shifts to the shoulders of women. Ascribing anomalous births to female agency would seem to take us right back to the notion of maternal impressions, which posited that a sin of commission would mark the unborn child. Today it is a sin of omission — failing to undergo prenatal testing — that marks the child, but the underlying belief that the woman is responsible for the outcome remains the same.

The diagnosis of FAS emerged in the United States during the baby bust of the 1970s, following the baby boom of 1946–64. In the 1960s and 1970s laypeople began to see reproduction as being within their mastery. Not only could parents now control the timing of pregnancy and birth, through vastly increased access to safe and effective contraception, they also could increasingly exercise control over the outcome of the pregnancy. "Yet, a subtle consequence of the success in preventing unwanted birth is that fertility now appears to be more within individual control."[14] The converse of greater personal control is greater personal responsibility. The increased attention to birth defects in this period exemplifies a societal shift in emphasis over the course of the twentieth century, as family size shrank, from "quantity" of children to "quality" of children.[15] Social pressure to have a "perfect child" complemented this demographic pattern.[16]

Concurrently, technological advances promised to make that goal attainable. New technology made it possible both to observe fetal development more closely and to treat fetal conditions such as rhesus (Rh) disease in utero; the fetus became a patient in its own right.[17] These changes precipitated a radical shift in the relationship between the fetus and the pregnant woman, both in her eyes and in the eyes of the public. As ultrasonography became widespread in the 1970s, doctors and the public could "see" the fetus in ways that had not been possible. The relationship between the pregnant woman and the fetus was thus reconfigured, and the relationship between the pregnant woman and society was transformed as well. The photo essay "The Drama of Life before Birth" by Swedish photographer Lennart Nilsson, published in *Life* magazine in 1965, exemplifies this perspective shift: the fetus literally became visible in a way it had never been before. Nilsson's photographs depicted in full color the stages of embryonic and fetal development, including the famous image of an eighteen-week-old fetus inside the amniotic sac that appeared on the cover of the magazine. (Ironically, all but one of Nilsson's pictures in this series were not taken in utero, as the accompanying text led readers to believe, but were pictures of aborted fetuses.)[18] Such imagery represented a dramatic change in the relationship not just between the pregnant woman and the fetus but between the public and the fetus. This kind of visualization had the effect of socializing pregnancy, of making the experience commonly accessible to all rather than a private, interior experience of the pregnant woman alone. As naturalistic images of fetuses began to appear in the public arena, the public began to feel it had a much greater stake in fetal welfare and consequently in women's actions during pregnancy. Pregnant women came to be viewed as a kind of public property. "Once this very private internal process became publicly visible the possibility arose for the public control of pregnancy."[19]

Prenatal diagnostic technology, particularly ultrasound, further eclipsed the woman in the act of illuminating the fetus.[20] The image produced by ultrasound literally reduces the woman and her body to background, while foregrounding the fetus. The pregnant woman appears to be nothing more than a vessel, a "fetal container."[21] The fetus came to be seen as "a fully formed 'preborn baby,' a free-floating being temporarily housed in the womb but with interests and needs of its own."[22] Elevating the status of the fetus had the consequence of devaluing the pregnant woman. In addition, ultrasound powerfully affected the way women experienced pregnancy. One doctor noted the way ultrasonography profoundly changed the relationship between a pregnant woman and the fetus developing in her uterus, as one of the doctors whom I interviewed described:

I really think that the single technological thing that happened, that changed all of that was ultrasound. Because any person who sees a baby sucking its thumb or kicking around, sees the heart beating, sees a three-dimensional living organism in the belly — and we [are] frequently doing these, you know, kinda 12 to 16 weeks, so we are still in the organ-forming, you know, time — gets a real shot in the head, you know, in terms of oh my God, you know, this thing is dependent on me. It's a real living organism and I am responsible for it . . . And now I think people expect perfect results because they have seen this kid and — but they also have a much more — there are interesting bonding studies. These people bond to this fetus, that is, you know, maybe it has no cortex . . . but they do. They see it and you know this makes it real to them, so therefore that sense of responsibility kicks in beyond themselves and they will do it for that fetus, what they would not do for themselves. [29F, female]

Recognizing this effect, some doctors even recommended using ultrasound as a way to persuade women to give up drinking or smoking. The irony of these technological changes is that even as they made the fetus more "real" to the pregnant woman, they also served to introduce a wedge between the woman and the fetus. As the woman "bonded" with the fetus, these technologies made it conceptually possible to think of the pregnant woman and the fetus as separate entities, thus setting the stage for the emergence of maternal-fetal conflict.

Reproduction and Control

The alienation of the pregnant woman and the fetus began in an era when individual women were gaining new levels of control over their fertility, through the introduction of oral contraceptives in 1965 and the legalization of abortion in 1973. Reproduction was rapidly politicized in the early 1970s, just as sexuality had been in the prior decade. Even as women struggled to make abortion a matter of individual conscience, reproductive matters shifted from the private realm to the public arena. At the very least, it is ironic that FAS was first diagnosed in 1973, the same year that abortion was legalized in the United States. Kristin Luker has noted the ways in which the debate about reproductive rights is deeply embedded in notions of women's social roles and responsibilities. The construction of FAS reflects the same conflicts "about women's contrasting obligations to themselves and others."[23] As with abortion, the discourse on fetal alcohol syndrome captures the sense that women cannot be trusted to act in ways once assumed to be "natural," intrinsic to their status as mothers or potential mothers.[24] This shift embodied not just social unease over the changing roles of women but social distress about women's growing propensity to act in ways that were independent of and perhaps in opposition to the maternal role as socially constructed. The 1960s and 1970s were a time of overt gender conflict and change, as women sought equality in the work-place, in higher education, and in the political arena, as well as in personal re-lationships. Women as individuals and as a group increasingly stepped out of and beyond their traditional roles. "A new social mythology encouraged the view that some women were not just bad mothers but 'anti-mothers,' who vio-lated their most fundamental natural instincts and who threatened to destroy the institution of motherhood altogether."[25]

At the same time, the feminist movement and the consumer health move-ment united powerfully in the women's health movement, which sought greater control and respect for women. The emergence of the women's health movement in the 1970s, exemplified by the publication of *Our Bodies, Ourselves* in 1973, directly challenged male orthodox medical authority and may also have contributed to a fraying of the bond between the (usually) male doc-tor and the female patient. The women's health movement highlighted the ex-tent to which women in the 1970s were fighting to change the rules of the game.

Even as women sought to change the rules and aspire to social roles out-side motherhood, the social construction of reproductive risk and particularly of fetal alcohol syndrome sought to keep women "in their place." A *JAMA* article on FAS, for example, proposed that to prevent the syndrome, "the no-tion of mothering from conception, not birth, must be fostered in the non-pregnant,"[26] as if to enmesh women further in a cult of motherhood. In this respect, concern about drinking before as well as during pregnancy was some-thing of a bellwether for the "preconception" movement that is gaining promi-nence today and that exhorts women to begin changing their diets, taking vi-tamin supplements, and avoiding alcohol, tobacco, and other substances when they are *considering* becoming pregnant. Thus, the ritual prohibitions of preg-nancy now extend to women who are no more than potentially pregnant — that is, all women of reproductive age. Indeed, the impulse to confine women to maternal roles may have been strengthened by the ascendance of feminism as a social movement in this period. Whether in the form of bra-burning or lobby-ing for passage of the Equal Rights Amendment, women directly confronted social stereotypes and demanded a new place in the social order. Society's re-newed emphasis on pregnancy served in part to remind women that repro-duction was their most valued social role.

Emergence of Maternal–Fetal Conflict

In this environment of sociocultural ferment and change, such cultural categories as motherhood and pregnancy easily metamorphosed and took on different shades of meaning. Most significantly, as the fetus gained in visibility and prominence, the bond between the pregnant woman and the fetus frayed considerably. The "organic unity of the fetus and the mother"[27] was challenged and upset by medical, political, legal, and social developments in this period, which was marked by an increasing tendency to see the pregnant woman and the fetus as being at odds, their individual welfares as oppositional rather than mutual. The social and gender dislocations of the time crystallized in the paradigm of maternal-fetal conflict. The idea of maternal-fetal conflict is predicated on the notion that the pregnant woman and the fetus are divisible; that as divisible and thus separate entities, they have individual as opposed to mutual interests, and that, furthermore, those individual interests may be in opposition to each other. The paradigm of maternal-fetal conflict is implicit in abortion. Its cultural ascendance is also evident in the rise in the number of court-ordered cesarean sections — cases in which women for personal or religious reasons refuse to consent to this procedure and hospitals or physicians respond by seeking judicial intervention to mandate it. These cases rest on the notion that the woman and the fetus do not have mutually intertwined interests but rather can be pitted against each other. Court-ordered cesarean sections also imply that the fetus must be "rescued" by physicians from the mother's body, highlighting the notion that women's bodies can be dangerous or threatening places for fetuses to be.

Maternal-fetal conflict also undergirds the wave of criminal prosecutions of pregnant substance users in the 1980s and 1990s. And maternal-fetal conflict colors our thinking about fetal alcohol syndrome and drinking during pregnancy: we conceive of harm accruing solely to the "innocent" fetus, perpetrated by a mother who is ignorant, selfish, or remiss. Conceiving of FAS as a manifestation of maternal-fetal conflict precipitates punitive measures such as incarcerating pregnant women who drink rather than providing them with health and social services (see publications by Dorris and Daniels for examples of these kinds of responses to FAS)[28] or attempting to expand the legal definition of child abuse to include drinking during pregnancy.

Individualizing Responsibility

The emergence of fetal alcohol syndrome as a medical and a social problem illustrates and reveals our tendency to see disease as the consequence of individual behavior, or what Sylvia Tesh has called "inappropriate lifestyles."[29] FAS illustrates an endemic problem of contemporary American society: the individualization or privatization of public health. Irving Zola noted that "the labels health and illness are remarkable 'depoliticizers' of an issue. By locating the source and the treatment of problems in an individual, other levels of intervention are effectively closed."[30] The work of the moral entrepreneurs who created the diagnosis of FAS and who disseminated it within the medical world illustrates the politics of assigning causality to disease: an undesirable birth outcome could be attributed to the moral failure of the mother who drank to excess during pregnancy rather than to any aspect of her environment, no matter how disadvantaged it may be, or to biology itself. Furthermore, the moral construction of the risks incurred by drinking during pregnancy individualized social responsibility for children's well-being. Thus, from the outset, the prevention of FAS was perceived to rest squarely with the individual and to be predicated on a change in her "lifestyle." The notion of "personal responsibility" that Ashley Montagu invoked in connection with thalidomide reverberates through medical writing on fetal alcohol syndrome. As one doctor wrote, "Unlike many other problems of newborns, FAS could easily be prevented by maternal avoidance of alcohol."[31] Even more bluntly, Sterling Clarren and David Smith wrote in the *New England Journal of Medicine,* "One fundamental fact should no longer be doubted in the medical or lay community: a large number of congenital malformations and central-nervous-system dysfunctions will be prevented through maternal avoidance of heavy liquor consumption during pregnancy."[32] To prevent fetal alcohol syndrome required only "maternal avoidance" — no collective action, no attempt to ameliorate social inequality, no concerted social change were necessary. This individualization of public health taps into deeply held American notions of free will, autonomy, and control over one's own destiny.

Moreover, in the eyes of the FAS entrepreneurs, women who drank during pregnancy delivered not only deformed babies but social problems as well. Recall the judgment voiced by Kenneth Lyon Jones and his colleagues at the

University of Washington in their first article on FAS: "the offspring of chronic alcoholic women . . . become a problem for society in postnatal life."[33] More recently, Ann Streissguth has written, "Drinking during pregnancy is linked with disaster — personal disaster for the 'bright-eyed ones,' disaster for their families and their hope for their young, and disaster for . . . society as we struggle to pay the huge price of this destruction."[34] According to this perspective, women who drink during pregnancy bestow destruction, disaster, and disorder both on their alcohol-affected infants and on society as well.

Fetal Alcohol Syndrome in the Public Arena

Just as social forces accelerated and magnified trends within the world of medicine, contributing to the emergence of fetal alcohol syndrome, they also had an impact on how the newly conceived social problem of drinking during pregnancy was perceived and the types of policy responses to it. Initially, there was scant attention to FAS outside the world of clinical medicine. As recognition of the diagnosis expanded beyond the boundaries of medicine into the public consciousness, drinking during pregnancy also broadened considerably as a social problem. Initial case reports identified FAS as a disorder associated not with heavy drinking alone but with the disease of alcoholism itself. (Recall that the mothers of the eleven children first described by the Seattle team were all severe, chronic alcoholics.) Over time, perception of the problem widened to include any and all drinking during pregnancy. The emphasis was no longer on women who were alcoholics but on all women.

A brief sketch of the history of policy responses to fetal alcohol syndrome at the national level reveals this fundamental shift. This policy history illustrates the essential characteristics of U.S. policy with respect to FAS. Policy is driven by the assumptions, first, that no safe level of alcohol consumption exists and, second, that complete abstinence from drinking is necessary.

The federal government's initial response to the new diagnosis did not occur until 1977, four years after the discovery of FAS. In that year, the National Institute on Alcohol Abuse and Alcoholism issued a statement that pregnant women who consumed more than six drinks a day were at risk of delivering babies with birth defects. In December of that year, the executive board of the American College of Obstetricians and Gynecologists (ACOG) also formulated a policy statement that, like the NIAAA's, focused on the dangers of heavy drinking.

Ample clinical and experimental evidence exists to demonstrate that chronic, excessive consumption of alcohol during pregnancy is associated with a variety of congenital malformations and other defects.

At the present time there is not sufficient evidence to substantiate or refute that moderate intake of alcohol is harmful to the fetus.

The use of intravenous alcohol to inhibit premature labor has not been implicated.[35]

The actions by both organizations reflect the initial construction of the diagnosis as a syndrome that occurred in the children of women whose drinking was "chronic [and] excessive." Nor is it surprising that these two organizations — one institutionally dedicated to alcohol abuse, the other to pregnancy and childbearing — were the first to make policy pronouncements about the new disorder.

However, by the early 1980s, the perception of FAS shifted, and doctors began to conceive of the risks posed by prenatal alcohol exposure much more broadly. No longer were doctors concerned solely with alcoholic women; now *all* pregnancies were deemed to be at risk from any amount of alcohol exposure. Following this new logic, Edward Brandt, the acting U.S. surgeon general, issued a warning in 1981: "The Surgeon General advises women who are pregnant (or considering pregnancy) not to drink alcoholic beverages and to be aware of the alcoholic content of food and drugs."[36] (This recommendation has not been changed since it was first issued.) The surgeon general's warning indicates a significant change in the locus of risk represented by prenatal alcohol consumption. This shift reflects not only evolving medical knowledge but a changing context for the interpretation of the new syndrome.

Just as, earlier in the twentieth century, more relaxed attitudes towards alcohol use had accompanied a shift in medical thinking about alcohol and reproduction, so in the 1970s, and particularly in the 1980s, concerns about alcohol and substance use began to resurface in American society. Alcohol consumption, and substance use more generally, began to reacquire the negative moral valence that advocates of the disease model of alcoholism had labored to remove. President Richard Nixon first declared a federal "War on Drugs" in 1972; President Ronald Reagan and First Lady Nancy Reagan reinvigorated that war a decade later. Nancy Reagan championed the "Just Say No" slogan. Mothers Against Drunk Driving (MADD), the populist social movement that sprung up in response to deaths from alcohol-related car crashes, was founded in 1980. The popular media heralded the country's "new temperance" and the "sobering of America."[37] In 1984, Congress passed legislation mandating a minimum drinking age of twenty-one, up from eighteen. President Reagan sought and passed a massive Anti-Drug Abuse Act in 1986. This reemergence of concern about alcohol and drugs signaled the birth of a latter-day temperance movement in the United States. American society began to focus on alcoholism not just as a disease that afflicted individuals but as a social scourge that could harm "innocent victims." And annual per capita alcohol consumption levels did fall in the United States in the 1980s, from 2.76 gallons of ethanol in 1980 to 2.42 gallons in 1989.[38]

Subsequent policy initiatives reflected this new conception of risk, particularly the assumption that there is no safe level of alcohol consumption during pregnancy and that, consequently, all women are at risk of having a baby with fetal alcohol syndrome. In 1983, in sharp contrast with ACOG's initial policy, the American Medical Association recommended that *all* women be screened for alcohol use as part of routine prenatal care. In 1984, Congress signaled its increasing attention to the problem of drinking during pregnancy by declaring the week of January 15 the first National Fetal Alcohol Syndrome Awareness Week. Four years later, Congress succeeded in passing the "label law," mandating that every bottle of beer, wine, or spirits sold in the United States be labeled with a warning about the risk of birth defects from drinking during pregnancy.[39] The label law went into effect in 1989, the same year that Michael Dorris published *The Broken Cord*. One year later, the first national nonprofit organization dedicated to fetal alcohol syndrome, the National Organization on Fetal Alcohol Syndrome, was founded.

In many ways, NOFAS exemplifies the expansion of risk associated with drinking. When NOFAS was founded, its executive director explained the need for the new organization by declaring, "I think a lot of middle-class and upper-class women don't know that occasional use of alcohol during pregnancy is dangerous."[40] In the decade since then, NOFAS has been dedicated both to helping families cope with children affected by FAS and to raising public awareness of the risks that it believes drinking poses. NOFAS has accomplished the latter goal in a number of ways. One of its first initiatives was to develop a specific medical school curriculum to educate medical students about drinking during pregnancy. William Kennedy Smith, who joined the NOFAS board of directors after his acquittal in the Palm Beach rape trial, was instrumental in getting his alma mater, the University of New Mexico Medical

School, to implement the NOFAS curriculum, the first school to do so. (Nine other schools have since adopted the NOFAS curriculum.) NOFAS has enlisted the aid of a number of celebrities, including the actors Jimmy Smits and Rodney Grant and the singers Bonnie Raitt and Queen Latifah, to appear in public service announcements and mass media campaigns that urge women to abstain from alcohol while pregnant. More recently, NOFAS has targeted adolescents and children as groups in need of more intensive education about the risks of drinking during pregnancy, by leading teen peer education workshops and by producing a book for kindergarten children to teach them "the importance of a pregnant woman's abstaining from alcohol."[41]

All these tactics, along with the surgeon general's warning and the alcoholic beverage label law, exemplify the broad-based and universalistic approaches of the American policy response to fetal alcohol syndrome. Each tactic is premised on the notion that any amount of alcohol is unsafe in every pregnancy. As the NOFAS slogan stipulates, "No safe time. No safe amount. No safe alcohol. Period." These policies are motivated by the belief that alcohol consumption is never safe in pregnancy and that complete abstinence from alcohol should therefore be required of pregnant women.

Media depictions of FAS have reinforced the notion that even a single drink could be dangerous and that women should feel morally obliged to abstain from alcohol while pregnant. Articles in magazines and newspapers bore headlines such as "An Innocent Inherits the Anguish of Alcohol" (Orlando Sentinel Tribune),[42] "Pregnancy, Alcohol Can Be Deadly Mix" (Minneapolis Star Tribune),[43] "Drinking Devastating to Unborn" (Seattle Times),[44] "Kids Pay for Prenatal Drinking" (USA Today),[45] "Children Pay the Ultimate Price for a Drink" (Houston Chronicle),[46] "Prescription for Tragedy: Alcohol and Pregnancy Stack Deck against Baby" (Seattle Times),[47] and, bringing us full circle, back to the nineteenth century, "The Tragic Inheritance: A Father's Chronicle of Fetal Alcohol Syndrome" (Washington Post).[48] Just as in the medical literature, the popular press stressed the "innocence" of the children with FAS, the "tragic" nature of the syndrome, and the alleged risk of even "a drink." These messages in the popular press, along with most policy responses, target all women as being at risk.

The Mismatch between Problem and Policy

Our policy responses imagine that the risk is diffuse and widespread, that the essential task is to inform and warn all women about the dangers of alcohol in pregnancy. Yet the small — though not negligible, to be sure — number of women and pregnancies that are truly at risk of fetal alcohol syndrome is concentrated, so the diffuse policy approach is inadequate to the task of actually preventing fetal alcohol syndrome. Most women and most pregnancies are *not* at risk. For those who *are* at risk — women whose own social and physical well-being is already compromised by their drinking — warning labels, point-of-purchase warnings, colorful posters, children's storybooks, television commercials, and the entire FAS "information, education, and communication" armamentarium are probably powerless to help. Broad-based policies target many women unnecessarily and they neglect the women actually at risk. Broad-based prevention efforts are not likely to succeed, because women who give birth to children with FAS are not simply random members of the general drinking population. A small proportion of women, especially those most disadvantaged by poverty and social isolation, bear the greatest burden of risk for FAS. Such policies do double damage by targeting the entire population of women unnecessarily while neglecting the women who lives are most adversely affected by drinking, before, during, and after pregnancy. In addition, such policies exacerbate our tendency to individualize blame. They give the illusion of concerted social effort to deal with the social problem, but without actually doing so in any meaningful way.

Moreover, many of our policy responses to fetal alcohol syndrome are punitive rather than preventive. Eleven states mandate that prenatal alcohol use be reported to child protection agencies (Illinois, Indiana, Iowa, Massachusetts, Michigan, Minnesota, Oklahoma, South Dakota, Utah, Virginia, and Wisconsin). Indiana, Wisconsin, and Utah define prenatal alcohol use as child abuse and neglect through statutory law. Both Wisconsin and South Dakota recently passed legislation to authorize civil detention or commitment for pregnant women who "lack self-control" in the use of alcohol.[49] Cases like Deborah Zimmerman's, in which women are prosecuted, or state laws that mandate reporting of prenatal alcohol abuse to child protection agencies or use statutory definitions of prenatal alcohol use as child abuse and neglect, are all aimed at punishing women *after* fetal damage has occurred. It is questionable whether such policies accomplish much in the way of prevention. Indeed, they do not really aim to prevent FAS so much as to punish women who fail to heed public health warnings. These punitive policies also raise issues of gender equity, violation of constitutional rights, and stigmatization of minorities, who are most often targeted under punitive laws. There is a regrettable dearth of evidence that any of these policies have done much to reduce the incidence of FAS. If anything FAS incidence appears to be rising, though our surveillance methods are simply not good enough to know one way or the other. There are certainly reasons to suspect policy failure. More than ten years of research on the alcohol warning label has failed to demonstrate any effect on people's drinking behavior. And indeed, the lack of findings with respect to the alcohol label merely echoes what we already know about warning labels generally. They do not work.

The United States seems exceptional when we view it in an international context. The worldwide incidence of fetal alcohol syndrome is estimated to be between 0.33 and 0.97 per 1,000 births.[50] The study that estimated this global incidence estimated the incidence in the United States to be 1.95 cases per 1,000 births and in Europe to be 0.08 cases per 1,000 births. (These estimates mask considerable differences within countries and regions. The incidence of FAS appears to be higher among severely disadvantaged populations everywhere, especially among American Indians in the United States, First Peoples in Canada, and blacks in South Africa.) Yet the United States does not have a higher alcohol consumption per capita than other countries. It ranks only thirty-second globally, with annual per capita consumption of 8.9 liters of pure alcohol, compared with 14.4 liters in South Korea, 13.7 liters in France, 11.7 liters in Germany, and 9.6 liters in Italy and Australia alike.[51]

The United States is not alone in recommending abstinence from alcohol during pregnancy; Australia, Austria, Canada, Denmark, and Sweden have similar warnings. Yet the U.S. surgeon general's advisory is stronger than comparable warnings in other nations. Both the United Kingdom and New Zealand have moderate warnings. For example, the British Royal College of Obstetricians and Gynaecologists' policy statement reads, "There is no conclusive evidence of adverse effects in either growth or IQ at levels of consumption below 120 gms (15 units) per week. Nonetheless it is recommended that women should be careful about alcohol consumption in pregnancy and limit this to no more than one standard drink per day."[52] Guidelines from Health Canada and from the Canadian Centre on Substance Abuse stress the importance of targeted rather than universal policies: "Resources should be directed toward pregnancies at high risk of FAS within high risk families."[53] Canadian, British, and European guidelines also stress explicitly the potential harm from policies or warnings that make FAS seem "inevitable even at low levels of consumption."[54] The World Health Organization stipulates that "It is not yet possible to definitely identify a safe limit of alcohol consumption during pregnancy. Moderate alcohol consumption during the first trimester of pregnancy (two or more drinks per week to two drinks per day) involves little if any additional risk of fetal malformation, but this might not apply for effects on intellectual development."[55] Most European governments, including France, Belgium, Germany, the Netherlands, Portugal, Spain, and Switzerland, have no official warning or recommendation on drinking during pregnancy.

Moreover, the construction of FAS in the American medical imagination stands in distinct contrast with that in Europe. In the United States the risk is construed as absolute and universal. Europeans, who drink at much higher rates than Americans and who have much more liberal attitudes towards alcohol, have long viewed the turmoil over FAS in the United States with bemused skepticism. The distinction between European and American attitudes is evident along several dimensions. First, Europeans take seriously the research question of whether there is in fact "a level of consumption which could be considered safe for the fetus which would allow the mother to take part in social gatherings, including meals, without absolute abstention."[56] It is all but impossible to imagine such a statement appearing in an American journal, infused as the American literature is with the spirit of temperance. The European literature is much more likely to *suggest* that women be "discouraged" from drinking during pregnancy, rather than to *mandate* that they abstain. In the United States, both risk and the subsequent prohibition against drinking are absolute; in Europe, the literature on FAS frequently acknowledges gradations of risk: "the evidence is consistent with there being no harm associated with the consumption of a drink a day or less during pregnancy."[57] Second, the European medical literature frequently evinces concern for the harmful effect of scaring women with dire caveats, noting that "strong warnings without scientific justification cause guilt and concern which may be harmful for the mother and her child."[58] Third, European researchers, unlike their American counterparts, whose writing betrays more moralism than pragmatism, have made a concerted effort to test interventions to reduce alcohol consumption during pregnancy among women who drink heavily. Finally, Europeans seem more willing to admit ambiguity and uncertainty, to acknowledge that "past experience warns against making global inference from limited scientific experience and mix[ing] scientific evidence with moral attitudes."[59] Although the European literature is not without its own moral overtones, the differences between American and European views of FAS parallel more general differences in risk perception between the two cultures.[60] The absolute abstention that characterizes American policy also reflects the litigiousness of American society: since a "safe" level is unknown, the only "safe" course is to eliminate risk by avoiding alcohol altogether.

The U.S. surgeon general's warning is rare in the strength of its recommendations that women should abstain from alcohol altogether and should, moreover, be vigilant about the miniscule alcoholic content of food and drugs. Nor is this dire warning without consequence. Women have called teratogen hotlines in a panic after eating rum raisin ice cream the day before they realized they were pregnant.[61] Even more striking is the congressionally mandated warning label on all alcoholic beverages sold in the United States. The United States is one of only nine countries in the world to mandate that a warning label appear on alcoholic beverages. It is the only developed country among the nine, and it is the only one to include a warning about alcohol and pregnancy on the label.

The kinds of policies that the United States has pursued to prevent fetal alcohol syndrome betray a particular American orientation towards social problems. The alcoholic beverage warning label is universal: it targets all women, pregnant or not, heavy drinker or not. But the label itself also reveals something about our societal priorities and our assignment of blame for the social harm from alcohol. The text of the label contains four distinct warnings (in order of their appearance): drinking during pregnancy; driving; operating machinery; personal health.

> GOVERNMENT WARNING: (1) According to the Surgeon General, women should not drink alcoholic beverages during pregnancy because of the risk of birth defects. (2) Consumption of alcoholic beverages impairs your ability to drive a car or operate machinery, and may cause health problems.

Drinking during pregnancy is set apart as the first and by implication the most significant or severe of the risks listed. It is also the most specific and detailed of the warnings, spelling out the consequences of birth defects. The final warning, about personal health, is couched in the vaguest terms — "may cause health problems," which remain unspecified. Only in the case of pregnancy does the warning label issue a direct injunction that "women should not drink alcoholic beverages." There is no similar admonition that "no one should drive a car or operate machinery" after drinking, for example.

While there is no denying that the public health consequences of alcohol use are severe in American society, the alcoholic beverage warning law is but a pale reflection of the impact of alcohol on American society. The NIAAA estimates that 100,000 deaths every year in the United States are attributable to alcohol.[62] These include more than 25,000 deaths from cirrhosis of the liver, one of the ten leading causes of death in this country,[63] and almost 16,000 deaths (and more than 300,000 injuries) in motor-vehicle crashes in which alcohol was a factor.[64] Fourteen million Americans meet diagnostic criteria for alcoholism or alcohol abuse,[65] and on any given day in the United States an estimated 700,000 people are in residential treatment for alcoholism.[66] The social harm from drinking and driving alone is staggering: an estimated 120 million episodes of alcohol-impaired driving occur every year. More than a million arrests are made every year for drunk driving, and 41 percent of all traffic fatalities involve a driver or pedestrian who was intoxicated.[67] The National Highway Traffic Safety Administration estimates that alcohol-related motor-vehicle crashes cost the nation $45 billion every year in direct costs, lost earnings, and lost productivity. Alcohol misuse is believed to play a role in one-half of all homicides, one-fourth of serious assaults, one-third of suicides, and one-fourth of rapes.[68] Drinking is also implicated in workplace accidents (alcohol use on the job results in estimated productivity losses of $119 billion annually),[69] barroom brawls, street riots, domestic violence (alcohol is implicated in 57 percent of domestic violence assaults by men against women and in 13 percent of child abuse incidents),[70] and unprotected sex and all its negative sequelae (a quarter of young adults report that they have had unprotected sex because of drinking or drug use).[71]

There are two ways in which the warning label seems to fall short. First, fetal alcohol syndrome is arguably not the most pressing social problem occasioned by alcohol use. Second, although drinking by women contributes to these adverse consequences, it is drinking by men that poses the greatest threats to personal and public health and the social order. The social harm caused by excessive alcohol use is much more likely to be caused by men. Alcohol use is more common among men, and men are more likely than women to become alcohol dependent. Men are more likely than women to drive after drinking and to be involved in fatal crashes. In 1996, for example, 84 percent of all drivers involved in alcohol-related fatal crashes were men.[72] The warning label projects a public message that it is women's drinking that is to blame for social disorder, that it is women whose access to alcohol must be controlled, monitored, and restricted.

Medicalization as a Moralizing Force

As I have demonstrated, the "discovery" and codification of fetal alcohol syndrome had a significant impact not only on medical knowledge and practice but on the social understanding and experience of pregnancy much more broadly. Does the construction of drinking during pregnancy as a significant risk thus represent the medicalization of yet another aspect of human experience? In many ways the trajectory of medical knowledge and response to FAS follows the path sociologists have described as "medicalization," but in other important ways medical conceptions of FAS and of prenatal harm from drinking stand in stark contrast to what we recognize as medicalization. The medicalization of FAS and drinking during pregnancy unfolds along similar lines to the medicalization of other social phenomena such as alcoholism, mental illness, addiction, and child abuse, but it has very different consequences.

Traditionally, medicalization has been viewed as an inherently humanistic enterprise, one that rescues undesirable aspects of human existence from the trash heap of moral stigma, transforming sin to sickness — or badness to sickness, in the words of Conrad and Schneider.[73] Thus, drunkenness became the disease of alcoholism, madness became mental illness, craving became addiction. In each of these instances, medicalization held out the promise of a more humanitarian response to human woes, offering technical expertise and the purported moral neutrality of the medical model as desirable "solutions" to social problems. This transformation of social deviance from a sin to a sickness has been heralded as one of the cardinal dimensions of the process of medicalization.

Yet health and illness are not morally neutral states in our society. As scholars such as Conrad, Riesman and Nathanson, and Rosenberg have shown, medicalization may simply substitute one kind of stigma for another.[74] Far from excusing women from culpability, the medical diagnosis of fetal alcohol syndrome only serves to increase the moral opprobrium of women who drink during pregnancy. Just as medicalization has been conceptualized to reduce moral stigma, it was also believed to reduce personal responsibility for one's condition. Illness or disease deserved treatment rather than shame. Unlike the "dope fiend" of the past, the addict of today cannot help himself. Yet here again, FAS is the exception that proves the rule: the medical diagnosis of FAS heightened rather than diminished the degree of personal responsibility for birth outcomes imputed to women. All our policy responses emphasize this in-dividualization of responsibility — from beverage warning labels and point-of-purchase signs to prosecutions of women who drink.

In such instances as alcoholism and mental illness, medicalization not only removes personal responsibility but typically leads to a restitutive treat-ment of deviance. As society moves towards a medical model of social control, as contrasted with a religious/moral model or a legal model, responses to de-viance, to social problems, typically become less punitive, more restorative. Medicine follows its traditional Parsonian role; that is to say, medicine func-tions to return people labeled as sick to their normal functions and roles in so-ciety, to enable individuals to resume social participation. Yet medicalization is neither restitutive nor restorative in the case of FAS. Rather, medicalization appears to lead us in the opposite direction, towards a more punitive stance, as Deborah Zimmerman's story so poignantly illustrates. Our conception of risk turns any drinking during pregnancy into a deeply deviant act, stripping women who drink of any vestige of their maternal role and excluding such women and their children from the social sphere, rejecting them as profoundly unworthy.

In other respects, however, the medicalization of fetal alcohol syndrome functions much as it does elsewhere, chiefly as a mechanism of depoliticiza-tion and an agent of social control. This element of medicalization operates powerfully in the case of FAS: the diagnosis profoundly individualizes the prob-lem of prenatal harm, erasing the contribution of such socially mediated influ-ences on the health of a newborn as environmental pollution, nutrition, even the social conditions in which the pregnant woman dwells — the quality of her housing, her work conditions, her medical care, her relationships with inti-mates. By zeroing in on an individual's drinking as *the* cause of undesirable birth outcomes, the medical conception of FAS obscures the heavy weight of social structure on patterns of well-being and disability in society. Because of this power to individualize problems (and thus solutions), medicalization removes problems like FAS from the public sphere, thus obscuring public responsibility.

Finally, by promulgating and legitimizing the diagnosis of FAS, medicine acts as an agent of social control. The existence of the diagnosis blurs the boundaries between deviant and nondeviant, problematic and nonproblematic, making any and all drinking by pregnant — or even potentially pregnant — women deviant and problematic. The medical construction of the "problem" of drinking during pregnancy in fact universalizes risk, making all women susceptible to prenatal harm from alcohol and thus equally subject to social control. So, even as medicalization seems to remove FAS from the public arena by individualizing risk and responsibility, it thrusts women more firmly into the domain of public control.

The moral dimensions of the medical classification of fetal alcohol syndrome are hardly unique. Our society's conceptions of disease are often weighted by moral valences as well as biological realities. However, the moral entrepreneurship that shaped the diagnosis of FAS was unusually strong, so much so that moral and emotional evidence sometimes overtook scientific fact. In a "world of limited and flawed knowledge,"[75] as our world must necessarily be, moral categories appeal for their simplicity and rigor. If doctors cannot say with any degree of certainty how much alcohol in pregnancy is safe, how much unsafe, they can invoke biblical authority where scientific expertise fails. Moral pronouncements are absolute; they admit no ambiguity, unlike the messy world of lived reality. As such, they are almost like magic bullets, quick fixes for otherwise intractable, insoluble problems.

The diagnosis of FAS is an expression not only of physical disorder — microcephaly, midfacial hypoplasia, vessel tortuosity, curvature of the spine — but of moral and social disorder as well. As the entrepreneurs who created the diagnosis of FAS gave witness to a clinical reality, they also bore witness to what they saw as social turmoil and moral disorder. Kenneth Lyon Jones, David W. Smith, Ann P. Streissguth, and their colleagues at the University of Washington saw evidence of a tragic disorder not only in their patients' bodies but in society as well — particularly in the lives of the alcoholic women who bore these children.

And yet there is a paradox at the heart of this act of moral entrepreneurship. The observers of FAS indeed identified an "evil which profoundly disturbed them,"[76] and accordingly they sought to extend the realm of their expertise and influence. They saw themselves as healers of society; yet their actual role as clinicians afforded them precious little capacity to "cure" either the women or the babies affected by heavy drinking during pregnancy. Medical doctors exerted their professional expertise to identify the new disease, yet their medical expertise granted them no more power than anyone else to prevent its occurrence or ameliorate its sequelae.

Most of the moral entrepreneurs who created the diagnosis of FAS retreated from the social problems that were at its root as being beyond their bailiwick. But in assuming that improved outcomes rest on individual change rather than broader change in the social conditions that put individuals at risk, these clinicians made FAS merely a victim-blaming strategy. As one doctor wrote, "Previous to this discovery, learning and developmental problems often found in children of alcoholics were attributed to a disruptive home life and poor caretaking."[77] In such sentiments there is almost a sense of relief that difficult, often intractable social conditions can be safely ignored and that poor outcomes can instead be attributed to a biological cause that can be ascribed to individual behavior rather than social conditions. Such an abdication of social responsibility may be particularly appealing for physicians, who, despite their moral entrepreneurship, are as a group known for their avowed reluctance to engage in the lives of their patients beyond the body itself.

Such entrepreneurship — self-aggrandizing, yet ultimately impotent — is typical of medicine, as Freidson notes. "The jurisdiction that medicine has established extends far wider than its demonstrable capacity to 'cure' . . . Thus, the medical profession has first claim to jurisdiction over the label of illness and anything to which it may be attached, irrespective of its capacity to deal with it effectively."[78] In fact, the moral force that doctors brought to bear on the diagnosis of fetal alcohol syndrome is directly related to their powerlessness as medical doctors to do much about it. What they could not cure as physicians, they hoped to banish as moralists.

CHAPTER 7

BEARING RESPONSIBILITY

Only connect.

—E. M. Forster

If we believe that "knowledge is the paramount social creation,"[1] then we must consider its power to bound our choices as a society. Even knowledge that is seemingly objective and empirical is socially constructed, as this book has shown. What determines the nature and shape of that construction? In part, knowledge is shaped by the answers we find to the questions we choose to ask. It is also shaped by the answers we want to find questions for and by the solutions that present themselves to us even before we are aware of just what the problem is. Our frame of inquiry may be individual or social, medical or moral, distal or proximate, particular or universal, but in choosing the frame, we help to determine what we will see as well as the avenues of action open to us. What we see in the diagnosis of fetal alcohol syndrome is a constellation of somatic and social stigmata that is determined by a specific and blameworthy individual behavior, namely, drinking during pregnancy. Thus, the current construction of fetal alcohol syndrome illustrates how we individualize both risk and responsibility for outcome. "This emphasis on individual responsibility may deny broader social responsibilities for health and disease."[2]

Moreover, the diagnosis of FAS illustrates how "as a natural symbol . . . the body in health offers a model of organic wholeness; the body in sickness offers a model of social disharmony, conflict and disintegration."[3] Fetal alcohol syndrome — a diagnosis applied to children who are quite literally disordered and disorderly — embodies social disharmony and conflict about changing gender roles and the disintegration of social responsibility.

That the contemporary construction of the relationship between alcohol and offspring has this symbolic dimension should alert us to one of the most powerful if invisible links between this construction and past knowledge about drinking and pregnancy. Regardless of whether turn-of-the-century doctors were observing in their patients what we today would diagnose as FAS, there are some significant similarities between the construction of medical knowledge then and now. First, in both eras doctors have been concerned with the way private, individual behaviors have public, social impacts. Thus, drinking during pregnancy is worthy of sustained attention since it affects not just the pregnant woman, not even just the pregnant woman and her fetus, but the whole of society as well. Second, because of those social consequences of individual behaviors, doctors have always seen pregnancy as worthy of special intervention. They have considered it their responsibility to protect future generations; they have acted both for the sake of the affected children and for the good of society. Third, doctors have used pregnancy as an opportunity to control women's behaviors and more generally to enforce particular gender roles. In both eras, women who fail to act "maternally" are held responsible for adverse outcomes. No less than today, doctors in the past individualized risk. Fourth, doctors both then and now drew connections between women's drinking and social disorder. Finally, the relationship between alcohol and offspring has always been somewhat contested. Doctors have engaged in debates about evidence and authority; the problem of alcohol and reproduction has always been a socially constructed one.

Perhaps the deepest parallel between late-nineteenth-century notions of alcohol and heredity and the contemporary diagnosis of FAS is the way in which knowledge — particularly medical knowledge, which enjoys a privileged status in American society — may function as a form of social control. In both the earlier and later periods, women's participation in the broader social world — and thus their abandonment of traditional roles and escape from the confinement of the domestic sphere — was rapidly expanding. Let me give but two examples from each era in two important realms: education and reproduction. The similarities between these two eras are suggested by the fact that many women's colleges, including Smith (founded in 1875), Bryn Mawr (1885), Vassar (1861), and Wellesley (1875), were founded in the last quarter of the nineteenth century, just as many elite educational institutions, including all the Ivy League colleges, finally accepted coeducation in the last quarter of the twentieth century. Thus, both periods witnessed a dramatic expansion in the educational opportunities open to women. Equally important, in both periods women exercised a greater degree of control over fertility than they had in the past: witness the "voluntary motherhood" movement of the late 1800s and the introduction of oral contraceptives in the 1960s. The expanded social opportunities for women in both periods precipitated a ritual retrenchment into traditional maternal roles.[4] Physicians, as arbiters of the social order, sought to construct women's primary responsibilities as maternal. By emphasizing the precariousness and potential danger of pregnancy, as well as women's foremost role in protecting the next generation, the contemporary construction of fetal alcohol syndrome, no less than the earlier temperance and eugenics movements, forces women to shoulder the burden and the blame for adverse outcomes and damaged offspring. Indeed, in both periods, much of the medical concern about the effect of alcohol on pregnancy has been focused not so much on the outcomes for individual women and children as on the repercussions for society. Whereas doctors in the late 1800s lamented "race suicide" and degeneration, doctors today worry that "a substantial proportion of the next generation is growing up without the ability to be functioning, contributing adults," as one pediatrician described her fears to me. In both periods physicians saw reproduction not as an individual, private act but as one that was ultimately social and public. And in both periods they marshaled medical knowledge to wield social control.

Pregnancy and pregnant women are particularly vulnerable to such ideological constraints and social control through medicine. Pregnancy epitomizes "factors of impossibility, and uncertainty in situations where there is a strong emotional interest in success."[5] As one obstetrician recalled, "I don't know why other doctors do obstetrics, but I think one of the things that appealed to me about obstetrics in the very beginning was the fact that it was such a pleasant experience. Very positive." Births are joyous events; we want and expect that pregnancies turn out well. This expectation of success combines with the enigma and unpredictability of pregnancy to make it a unique phenomenon. Pregnancy transcends the biophysical processes of conception and gestation; something magical happens in the creation of new life. It is a process both very serious and dangerous, as well as vulnerable to danger.

As we have come to understand more about the physiology of pregnancy, we have also placed much greater emphasis on controlling the process. Thus, pregnant women are increasingly subject to public monitoring and overt social control. What had been a private and highly individual experience has become a public one, subject to surveillance and oversight by strangers. Over the course of the twentieth century, the locus of danger in pregnancy has shifted significantly, from the pregnant woman to the fetus. Today, women in the United States are largely safe from the devastating morbidity and mortality that accompanied pregnancy and childbirth in the past. As threats to women's lives have receded, the fetus—the new life at stake in pregnancy—has come to seem ever more imperiled and vulnerable. In part, the perception of an increasing threat to the fetus has enabled doctors to extend their monitoring and control of pregnancy and birth beyond the examination room, beyond the labor and delivery room, into their patients' private lives. An inherent tension has thus entered the maternal-fetal relationship. The greater the protections the fetus is offered, the greater the encroachments on women's autonomy, both bodily and spiritually. The greater the freedoms offered to women, the greater the purported threats to the fetus. Traditionally, it has been medicine that has monitored this relationship. Increasingly, it is society more broadly, as well as the state more directly, that polices pregnancy. The emphasis on maternal-fetal conflict leads us to take a criminal justice approach rather than an approach grounded in public health. The diagnosis of FAS, and more generally the societal response to drinking during pregnancy, is another manifestation of this notion of maternal-fetal conflict. After all, what are the "pregnancy police" protecting? They are policing the pregnant woman's choices and actions and protecting the fetus against harms that may be inflicted by the mother.

Yet thinking in terms other than maternal-fetal conflict might enable us to craft societal responses that offer more hope of helping women and children rather than policing and punishing. This extension of control has been "successful" in the sense that many women have internalized medical values and admonitions—sometimes even to the extent that doctors are scornful of lay "hysteria" about overimagined risk. At the same time, this extension of control has been imperfect in that many women remain truly at risk of adverse outcomes, not only because they have not been socialized to fear pregnancy but because both medicine and society at large have abandoned attempts to reach them or to change the conditions of their lives that predispose them to risk.

One of the goals of knowledge is to impose order on the chaos of human experience, "the raw confusion of the real world."[6] However, knowledge, once constructed, can be used to obscure certain dimensions of that real world even as it illuminates others. Since we "know" FAS is caused by fetal exposure to alcohol when a woman drinks during pregnancy, we cannot recognize the ways in which individual lives are blighted and hurt by social inequality. Just as the profession of medicine, in response to shifts in wider social values, became blind to its own past knowledge, "forgetting" in the mid twentieth century the answers it had found to the question of alcohol and offspring, we are blind to the ways in which "medicalization inevitably entails a missed identification between the individual and the social bodies, and a tendency to transform the social into the biological."[7]

That is, we choose not to see how deeply intertwined are individual "choices" in social circumstances. For example, many pregnant women who drink heavily are victims of violence and domestic abuse.[8] Women do not just "take" risks voluntarily; they live lives at risk, in circumstances that are often beyond their control. Drinking during pregnancy is stratified by race, education, and social class, as we saw in Chapter 5. As Brandt notes with regard to cigarette smoking, "to emphasize individual accountability is to deny that some groups may be more susceptible to certain behavioral risks, that the behavior itself is *not* simply a matter of choice" (emphasis in original).[9] Risk is not randomly distributed throughout the population.

Nevertheless, the paradigm of personal risk is ascendant in health today. We have transformed chance into risk, risk into danger, randomness into quantifiable risk factors. I have argued that attitudes about risk are a way of coping with uncertainty, of dealing with the cruel randomness of nature and of life. I now want to suggest that we should not necessarily think of outcomes like fetal alcohol syndrome as random but rather as deeply socially situated. Many adverse birth outcomes remain random; we have not yet deciphered the pattern or the predictors. Yet in other instances the precursors of unwelcome birth outcomes may seem clear, if unacknowledged. How we assign risk and responsibility for illness depends in part on what we see as the cause of illness. Is the disorder we diagnose as fetal alcohol syndrome caused by exposure to alcohol in utero, or is it caused by poverty, social distress, and the deeply rooted inequality of American society? "Simply identifying individual behavior as the primary vehicle of risk negates the fact that behavior itself is, at times, beyond the scope of individual agency. Behavior is shaped by powerful currents — cultural, psychological, as well as biological processes — not all immediately within the control of the individual. Behaviors such as cigarette smoking are sociocultural phenomena, not merely individual, or necessarily rational."[10]

Public health theory and practice have long held that social conditions can be forceful determinants of health, disease, and death.[11] Whether or not we believe that social conditions cause disease is a question not of biology but of moral evaluation. How we assign etiology to health and illness reflects our sense of the proper social order, no less than our understanding of particular biological processes.

It should be a matter of no small mystery to us that even most women who drink heavily during pregnancy give birth to children who are unmarked by the stigmata of fetal alcohol syndrome.[12] And it should be a matter of no small concern to us that most babies with FAS are born to women in particular and constrained social circumstances.[13] Yet we universalize risk in pregnancy and the specter of poor outcome: according to the official prohibitions of the U.S. surgeon general, for example, *all* pregnant women who drink are at risk of having a baby with fetal alcohol syndrome.

I have suggested thus far that there is a larger latent agenda in the way our society understands fetal alcohol syndrome and the relationship between alcohol and offspring. First, our constructions of risk in pregnancy function to preserve a particular social order and set of institutionalized gender roles and relationships. Second, our understanding of the etiology of FAS denies collective social responsibility for future generations by individualizing blame. Third, this understanding perpetuates social inequality by ignoring the poverty, chaos, suffering, and insufficiency of some women's lives that may be the truer causes of adverse outcomes for babies and children. Fourth, our conceptions of the risks of women's drinking, centered as they are around pregnancy, dismiss the harm done to women themselves by drinking and the social suffering that so often precipitates it. Like the current debate about "secondhand smoking," the construction of fetal alcohol syndrome is centered around the threat of injury not to the self but to the other. That bias is particularly evident in the warning label on alcohol bottles: unlike the labels on cigarette packages, the label contains only the vaguest warning about the extensive personal morbidity and mortality that are caused by drinking.

Finally, our preoccupation with drinking during pregnancy reflects a subtle but significant displacement of risk and responsibility. As noted by the doctors I interviewed, drinking does in fact impose a tremendous burden of morbidity and mortality on American society; however, the share of that burden caused by men's drinking—which is related to extensive personal morbidity, years of productive life lost, premature death, work absenteeism, violence (including violence against women), assault, crime, and motor-vehicle crashes— is vastly greater than that caused by women's drinking. I do not mean to suggest that women's drinking is without social consequence; but in American society men drink at much higher rates than women and their drinking has far more severe social consequences. That we nevertheless evince so much concern about fetal alcohol syndrome and drinking during pregnancy suggests that we as a society mislocate responsibility not only for individual cases of ill health but for social ills as well. I have already mentioned the peculiar ordering of the warnings on alcohol labels: the admonition about drinking during pregnancy precedes that about drinking and driving. Our policy responses both individualize responsibility and displace that individual responsibility onto one gender only. Women, by virtue of the biological capacity of their bodies, are made to bear responsibility for society. This sentiment appears as strong today as it was a hundred years ago, when the Anti-Saloon League exhorted women, "You can't drink liquor and have strong babies . . . For my own, my brother's, my country's sake, I will be a total abstainer."

Thus far I have written about the ways in which we as a society generally think about risk in pregnancy, about the social disorder didactically symbolized in fetal alcohol syndrome, about responsibility for individual health and social welfare. Let me conclude by addressing medicine's particular role in these matters. By diagnosing children with FAS and assigning a one-dimensional etiology — exposure to alcohol in utero — for the disorder, medicine further obscures social problems and the multidimensional causes of birth outcomes that are less than optimal. (Let me point out that this reductionist approach is typical of our market-driven society more generally.) The irony is that this individualistic approach to treating the physical disorder of fetal alcohol syndrome will never cure either the individual cases of FAS or the social disorder that is at the root of the syndrome.

How might we close the gap between the uncertainty of our medical knowledge about the relationship between alcohol and reproduction and the need to take action at a societal level to prevent adverse birth outcomes? Let us return to the story of Deborah Zimmerman, the Wisconsin woman who faced charges of attempted murder because she drank while pregnant. If we are attentive to some of the features of this case, we may begin to see some ways to rethink our notions of risk and responsibility and to extract from this sad story some principles that might guide us towards better public policy in this realm.

Deborah Zimmerman was thirty-five years old in 1996, when she gave birth to her daughter. She herself was the daughter of an alcoholic father, and she had had a drinking problem since high school — in other words, for more than half of her life. She had been raped three times and had been the victim of domestic violence in an earlier relationship — both experiences that can trigger or exacerbate problematic substance use among women. (Among women who are alcohol or drug dependent, 74 percent report sexual abuse in their past and 70 percent report current or past domestic violence.)[14] Moreover, Zimmerman's drinking had gotten her into trouble before; she had spent time in jail on a vehicular manslaughter charge after she killed a man while driving drunk in 1983.[15] And she had at least one prior suicide attempt. Indeed, despite the district attorney's contention that Zimmerman was attempting to commit homicide that March night in the emergency room, she was in fact threatening to commit suicide. She wanted to take her own life. In her own words, "*I'm going to drink myself to death.*" The court explicitly recognized this fact in its opinion, when it wrote, "Any intent or indifference that she may have manifested by her continued dependence on, and abuse of, alcohol during her pregnancy was directed toward her own body and the unborn child she carried within her, not toward another human being."[16] Thus, the court implicitly rejected the notion of maternal-fetal conflict in the Zimmerman case. Zimmerman's drinking had posed a threat to herself and to others around her long before she became pregnant.

Yet our conceptions of the risks of women's drinking, centered as they are around pregnancy, dismiss the harm done to women themselves by drinking, as well as the social suffering that so often precipitates heavy drinking. It is as if a woman's drinking problems have no social meaning until she becomes pregnant. Zimmerman's story suggests that we ought to focus on problem drinking behavior in women long before they get pregnant, for their own sake as well as for the sake of any future children they may have. How different might Deborah Zimmerman's life have been if she had received care for her drinking problem before Megan was even conceived?

The matter of Megan's conception provides yet another insight. Deborah Zimmerman claimed not to have wanted her pregnancy. In that regard, she was not unlike many American women. Surveys reveal that half of all pregnancies in the United States are unwanted or mistimed. In the National Maternal and Infant Health Survey examined in Chapter 5, women who reported their pregnancies were unwanted were twice as likely to drink at high risk levels as women whose pregnancies were wanted. Moreover, women whose pregnancies are unintended are less likely to recognize that they are pregnant early in the first trimester. These facts suggest that we ought to reorient our efforts to "prevent" fetal alcohol syndrome. Rather than striving to reduce drinking among women who are pregnant—an especially formidable goal for women who are dependent on alcohol — we ought to increase our efforts to reduce the incidence of unintended and unwanted pregnancies among all women, especially those who drink at high risk levels. Targeting unintended pregnancy rather than drinking is a broad-based policy, one that is warranted by the real magnitude of the problem that it seeks to address.

Deborah Zimmerman's case also highlights the futility of the alcoholic beverage warning label as a societal response. It strains credulity to believe that a warning label might have had any effect on her behavior. Rather than pursuing such universalistic policies as warnings, we should target the drinking of women like Zimmerman, whose alcoholism poses the greatest threat to themselves as well as to others.

Looking harder at Deborah Zimmerman's life also forces us to reconsider accountability, the issue that mattered so much to Joan Korb, the prosecuting attorney. Who is accountable for birth outcomes? Women? Families? Doctors? Social institutions? Society? If we believe our own rhetoric about pregnancy and childbearing as the future of society, then all of us, ultimately, must be held accountable for ensuring that women have the opportunity and resources to give birth to healthy babies.

Moreover, the stresses and strains of the collective body politic that are expressed in individual flesh-and-bone bodies are not readily amenable to care or cure by physicians, who treat the frailties and heal the injuries of particular human bodies. Yet in their patients, doctors inevitably encounter the weaknesses and sicknesses of the social body. How do they reconcile the relationship between these two kinds of body, these two kinds of illness, the individual and the social? The dilemma presents a paradox of care: often doctors cannot heal the individual body without treating the social one as well, yet to do so is well beyond their professional training and mandate.

In pregnancy — which by definition involves two lives, and the present as well as the future — this paradox of care is particularly keen. For many doctors, the social dimensions of FAS are as real and as acute as the somatic ones. They seem as incurable as well. Doctors have always viewed themselves as society's healers; they cure what ails us. Yet they must focus their therapeutic efforts on individual patients, as if to heal society one body at a time. Thus medicine emphasizes prevention at the individual level rather than reform at the societal level. By concentrating their efforts on prevention, physicians reveal their need to assert control over unpredictability as well as medicine's powerlessness to attack the broad social inequities that are manifest in their patients' ill health.

The "one interrelated unit" that a family practitioner spoke of in Chapter 4 is not just the pregnant woman and the fetus — the promise of new life she holds within her — but also a metaphor for the pregnant woman and the rest of us, society as a whole. The individual and the social are inextricably bound together. Indeed, the interrelationship of public and private is the fulcrum around which sociology turns: it is where we sociologists seek to uncover meanings.

We can look at this interrelationship of individual and collective, of private and public, as a mandate to police and control the behavior of individual women. Or we can regard the interdependence of individual and society as a moral imperative to ensure health for our entire population. That medicine as a discipline has at times chosen the latter path is a credit to its status as a healing profession. That it too often does not do so today, but rather focuses narrowly on individual risk, "lifestyle," and behavior, suggests how impoverished our sense of mutual obligation has become, how thoroughly we have abandoned the hope of collective responsibility for social welfare. No matter how we conceive risk, in the end all of us ought to bear responsibility for the present and future well-being of our society and especially for those least equipped to safeguard their own health.

DATA AND METHODOLOGY
FOR CHAPTER 5 ANALYSES

Description of Sample

The 1988 National Maternal and Infant Health Survey is the source of the data used in this analysis. The NMIHS was a nationally representative survey of women between the ages of fifteen and forty-nine who experienced a pregnancy in 1988. The NMIHS drew a stratified, systematic sample based on 1988 vital records of fetal deaths, live births, and infant deaths from forty-eight states, the District of Columbia, and New York City.[1] Black women and women delivering a low-birthweight baby were oversampled. The survey was administered by mail, phone, and personal interview, with a response rate of 71 percent.[2] The NMIHS provides data on 9,953 pregnancies. See Table A.1 for a summary description of the sample.

Due to the oversampling, blacks compose almost half (48 percent) of the total observations. White women account for about 49 percent of the sample, and Asian Americans, Pacific Islanders, and American Indians together account for about 3.5 percent.

Fifty-five percent of the women were married at the time of the survey; almost 40 percent were unmarried. A quarter of the sample had not completed high school, 40

---

1. South Dakota and Montana were not included in the sampling frame. According to the 1991 BRFSS, these two states are among those with the highest rates of prenatal drinking. Centers for Disease Control, "Frequent alcohol consumption among women of childbearing age — Behavioral Risk Factor Surveillance System, 1991," *Morbidity and Mortality Weekly Report* 43, no. 18 (1994): 328–29, 335.

2. U.S. Department of Health and Human Services, National Center for Health Statistics, *Public Use Data Tape Documentation, 1988 National Maternal and Infant Health Survey* (Hyattsville, Md.: U.S. Department of Health and Human Services, National Center for Health Statistics, 1991).

Table A.1. Sociodemographic Characteristics of the Sample of 9,953 Women in the 1988 National Maternal and Infant Health Survey (NMIHS)

Characteristic	Percent
Race	
White	48.9
Black	47.6
Other	3.5
Marital status	
Married	55.0
Unmarried	39.5
Missing	5.5
Years of education	
<12	24.5
12	39.9
13–16	31.7
≥17	3.9
Age at birth	
<20	16.6
20–29	54.3
30–34	17.8
≥35	7.6
Missing	3.8
Children ever born[a]	
1	30.6
2	26.6
≥3	38.2
Missing	3.2
Household income[b]	
<$10,000	33.5
$10,000–19,999	23.1
$20,000–34,999	20.8
≥$35,000	22.5
Pregnancy intentions	
Wanted	50.6
Mistimed	40.0
Unwanted	9.4

[a]Including the survey child.
[b]During the 12 months before the birth of the survey child.

percent were high school graduates, 32 percent had attended college, and 4 percent had graduate or professional training.

Teen mothers compose 17 percent of the sample; 54 percent of the women were in their twenties at the time of the delivery; 18 percent were thirty to thirty-four years old; and 8 percent were thirty-five or older. Almost a third of the women in the NMIHS were giving birth for the first time; 27 percent were experiencing a second birth, and 38 percent a third or higher-order birth.

One-third of the women in the sample reported household incomes of less than $10,000 in the year before the birth. Twenty-three percent reported incomes between $10,000 and $20,000; 21 percent, incomes between $20,000 and $35,000; and 23 percent, incomes of at least $35,000.

The proportions of women who drink at each risk level were calculated both for the full sample of all women and for the subsample of women who reported drinking before pregnancy. These proportions are presented in Table A.2.

I used multiple logistic regression analysis, which is appropriate for use when the dependent variable is dichotomous. I estimated a series of models concerning drinking behavior before and during pregnancy (results are presented in Table A.3). The first three models are estimated using the entire sample of 9,953 women. The dependent variables are whether the woman reported drinking before pregnancy (model 1), whether the woman drank at high risk levels before pregnancy (model 2), and whether the woman drank during pregnancy (model 3). The next two models were estimated using the subsample of 3,381 women who reported drinking before pregnancy. The dependent variables were whether the woman reduced drinking during pregnancy (model 4) and whether the woman reported drinking at high risk levels while pregnant (model 5). In addition, I used multiple logistic regression analysis to predict the odds of receiving late or no prenatal care and the odds of receiving medical advice on drinking during pregnancy.

The independent variables in all models are treated categorically. White is used

Table A.2. Reported Drinking Levels before and during Pregnancy, 1988 NMIHS

| | Before Pregnancy | | During Pregnancy | |
| | All Women (N = 9,953) | Women Who Reported Drinking before Pregnancy (N = 3,881) | All Women (N = 9,953) | Women Who Reported Drinking during Pregnancy (N = 3,881) |
Drinking Level				
None	61.4%	—	83.1%	56.4%
<3 drinks/wk	28.6	73.8%	14.6	37.7
3–13 drinks/wk	8.5	22.0	1.9	4.8
14+ drinks/wk	1.4	3.7	0.4	1.1

Table A.3. Odds of Drinking before and during Pregnancy, Estimated Using the 1988 NMIHS

Characteristic (reference category)	1 Odds of Reporting Pre-pregnancy Drinking (N = 9,953)	2 Odds of Drinking at High Risk Levels Pre-pregnancy (N = 9,953)	3 Odds of Drinking during Pregnancy (N = 9,953)	4 Odds of Reducing Drinking during Pregnancy (N = 3,881)	5 Odds of Drinking at High Risk Levels during Pregnancy (N = 3,881)
Race (white)					
Black	0.45**	0.58**	0.65**	0.70**	2.04**
Other	0.34**	0.37	0.59**	1.19	1.41
Years of education (12)					
<12	0.82**	1.34	0.96	0.77*	1.48*
13–16	1.33**	0.78	1.41**	1.32**	0.49**
17+	2.16**	0.49	1.62**	2.14**	0.96
Age (20–29)					
<20	0.58**	0.48**	0.68**	0.84	0.45**
30–34	1.1*	1.47*	1.27**	0.79**	1.57*
35+	0.86**	1.55*	0.93	0.62**	1.69*
Parity (1)					
2	0.96	1.07	1.28**	0.83	1.49
3+	1.03	1.23	1.53**	0.64**	2.35**
Unmarried (married)	1.32**	2.49	1.15**	1.06	1.10
Income ($35,000+)					
<$10,000	0.54**	1.16	0.92	0.80	3.11**
$10,000–19,999	0.54**	1.10	0.73**	1.05	2.25**
$20,000–34,999	0.62**	0.92	0.81*	0.90	1.62
Pregnancy intention (wanted)					
Mistimed	1.24**				
Unwanted	1.16*				
Pre-pregnancy drinking level (low [<3/wk] or none)					
Medium risk (3–13/wk)			12.82**	1.62**	
High risk (14+/wk)			25.35**	0.97	
Prenatal care (1st trimester)					
2d trimester			1.09	1.01	1.94**
Late or none			2.51**	0.81	1.29
Advised to reduce drinking (no advice)			1.38**	1.48**	1.44

*$p < 0.05$
**$p < 0.01$

as the reference category for race; the "other" category is composed primarily of Asian American and American Indian women. Educational attainment is analyzed using the years of schooling the woman has completed; twelve years is the reference category. Women aged twenty to twenty-nine at the time of delivery form the reference category for age. Parity is considered using three categories; in each case, the index child is included in the calculation of children ever born. Women experiencing their first birth form the reference category. Marital status is treated dichotomously, with married women forming the reference category. Household income, a proxy for economic status, is divided into four categories: less than $10,000, $10,000 to $19,999, $20,000 to $34,999, and $35,000 or more. The last category is the reference. Although household income is an imperfect measure of a woman's economic status, as it does not control for the size, age, or sex composition of the household, it is the best measure available in this data set. In addition, I created two sets of interaction variables. The first consisted of interactions between parity and age. The second consisted of interactions of black race with age, education, parity, marital status, and income.

Finally, some models control for pre-pregnancy drinking levels, timing of prenatal care, and receipt of medical advice on drinking. Women's drinking levels are divided into three levels of risk, according to the number of drinks consumed per week. One shortcoming of the NMIHS is that it does not provide any information on the amount of alcohol consumed per drinking episode. Thus, a woman who reported having 5 drinks per week may have had one drink every night or may have had a drinking binge one night a week. The high risk category includes women who reported drinking 14 or more drinks per week; the medium risk category, women who drank 3 to 13 drinks per week; and the low risk category, women who drank fewer than 3 drinks per week and perhaps as few as 1 drink per month. The low risk category is used as the reference category. The prenatal care variable is defined according to the timing of the woman's first prenatal visit. The reference category consists of women who received prenatal care in the first trimester. Women reported receiving different kinds of medical advice during pregnancy. Almost all women reported being given dietary advice, but smaller proportions reported being advised to stop smoking or drinking (see Table A.4). A dichotomous variable indicates whether the woman reported receiving medical advice on drinking during pregnancy; no advice serves as the reference category.

Table A.4. Women Receiving Medical Advice during Pregnancy, 1988 NMIHS

Type of Advice	Percentage of Women Reporting
Take vitamins and minerals	91.5
Eat proper food	86.1
Reduce smoking	61.9
Reduce drug use	59.7
Reduce drinking	58.6
Breastfeed	41.7

Logistic Regression Results

Model 1 presents the odds of drinking before pregnancy among all women in the sample (see Table A.3). Black women are less than half as likely as white women to drink before pregnancy. Other minorities (primarily Asian Americans and American Indians) are only two-thirds as likely as white women to drink. The odds of drinking increase with education and are more than twice as high among women with a graduate degree as among women who have not completed high school; the latter are about 20 percent less likely to drink than high school graduates. The odds of drinking are highest among women aged thirty to thirty-four and lowest among teenagers. Women with a higher parity are about 14 percent more likely to drink before pregnancy than women with only one child. However, controlling for pregnancy intention reduces the effect of parity and age. Women whose pregnancies were unintended have slightly higher odds of drinking before pregnancy. Unmarried women are 40 percent more likely to drink than married women. The odds of drinking increase with income.

Model 2 presents the odds of drinking at high risk levels before pregnancy. For the most part, the findings of this model mirror those of model 1, suggesting that the correlates of high-risk drinking before pregnancy are similar to the correlates of any drinking. The odds of high-risk drinking are lower for black women and for women less than 20 years old. There are no significant differences by parity, marital status, or income.

Model 3 presents the odds of drinking during pregnancy. In addition to the set of sociodemographic maternal characteristics, this model introduces controls for a woman's level of pre-pregnancy drinking. The model demonstrates that the correlates of drinking during pregnancy are, not surprisingly, quite similar to those of drinking before pregnancy. As with pre-pregnancy drinking, the odds of drinking during pregnancy are significantly lower for blacks and other minorities than for whites. Black women are only two-thirds as likely as white women to drink during pregnancy. The odds of drinking increase with education. Women who have attended college are 41 percent more likely to drink during pregnancy than women with twelve years of education, and women with graduate training are 62 percent more likely to do so. The odds of drinking while pregnant are 32 percent lower among teenage mothers than among women aged twenty to twenty-nine; however, they are 27 percent higher among women aged thirty to thirty-four. There is a strong parity effect, with each additional child increasing the odds of drinking by about 25 percent. Unmarried women are 15 percent more likely to drink than married women. The odds of drinking during pregnancy are lower among women with family incomes less than $35,000; however, they rise with income. As in model 1, black women with less than twelve years of education are almost twice as likely to drink during pregnancy, and black unmarried women and black mothers with a third or higher-order birth are each about a third more likely to drink during pregnancy.

A woman's pre-pregnancy drinking level is strongly predictive of the odds that she will drink during pregnancy. As might be expected, women who consume more drinks before pregnancy are substantially more likely to drink during pregnancy than women who consume few drinks or do not drink at all. Women who have 3 to 13 drinks per week before pregnancy are twelve times more likely to drink during pregnancy than women who have fewer than 3 drinks per week or do not drink at all, and women who have 14 or more drinks per week are twenty-four times more likely to do so. Receiving late or no prenatal care is associated with higher odds of drinking during pregnancy. Perhaps unexpectedly, receiving medical advice to reduce drinking is associated with 38 percent higher odds of drinking during pregnancy. However, this result can probably be explained by reverse causation; that is, women who drink during pregnancy are more likely to receive advice to reduce alcohol consumption than women who do not drink.

The final two models are estimated using only the subsample of women who reported drinking before pregnancy. Model 4 predicts the odds that drinkers will reduce alcohol consumption during pregnancy, controlling for pre-pregnancy drinking levels and prenatal care and advice. Model 1 suggests that black women are less likely to drink than white women; model 4 suggests that black women who do drink are less likely than white women who drink to reduce drinking during pregnancy. The odds of reducing drinking increase sharply with educational attainment. Thus, although women with more education are more likely to drink before pregnancy, they are also more likely to reduce alcohol consumption while pregnant. Women with less than twelve years of education are 23 percent less likely than high school graduates to reduce consumption, but women who have attended college are 32 percent more likely and women with graduate education are more than twice as likely to reduce. The odds of reducing alcohol consumption fall as age rises; women aged thirty-five and older are 40 percent less likely to reduce than women twenty to twenty-nine years old. Higher parity women are also less likely, by about one-third, to reduce drinking than women experiencing their first birth. There are no significant differences in the odds of reducing drinking by marital status or family income, despite differences in the odds of drinking before pregnancy within these categories.

Women who drank moderately before pregnancy (3 to 13 drinks per week) are 62 percent more likely to reduce alcohol consumption during pregnancy than women who had fewer than 3 drinks per week before pregnancy. These higher odds might be expected, since the higher a woman's initial drinking level, the greater the need for her to reduce drinking during pregnancy to avoid the risk of adverse birth outcomes. However, among women who drank at the highest levels before pregnancy (more than 14 drinks per week), the odds of reducing drinking while pregnant are no greater than among women who drank at low levels. Receiving prenatal care and the timing of such care do not affect the odds of reducing drinking. However, receiving medical advice to reduce alcohol consumption raises the odds of doing so by almost 50 percent.

The final model predicts the odds of drinking at high risk levels (14 or more drinks per week) during pregnancy. In many respects, model 5 is the most important model, since it provides evidence of the correlates that place women at risk of giving birth to a child affected by FAS. In this model, there are some sizable reversals in the odds associated with various characteristics. For example, black women are less likely than white women to drink before or during pregnancy (see models 1 and 3) and less likely to drink at high risk levels before pregnancy (model 2). However, their odds of drinking at high risk levels during pregnancy are twice those of white women. Similarly, less-educated women have higher odds of high-risk drinking during pregnancy despite their lower odds of drinking overall, and women who have attended college are only half as likely as high school graduates to drink at high risk levels, despite their higher odds of drinking before pregnancy. The odds of drinking at high risk levels while pregnant increase with age and parity in a pattern similar to that observed for drinking before pregnancy (compare models 1 and 5). In fact, women aged thirty-five and older are almost 70 percent more likely to drink at high risk levels than women aged twenty to twenty-nine, and women experiencing their third or higher-order birth are more than twice as likely to drink at high risk levels as women experiencing their first birth. There are no significant interactions of parity and age in this model. However, black women aged thirty-five and older are 60 percent less likely than white women to drink at high risk levels. The odds of high-risk prenatal drinking are higher among lower income levels than among women with family incomes of $35,000 or more; however, the odds fall as income rises.

Differences in Receipt of Prenatal Care and Advice on Drinking

To date, the two main modes of informing women of the risks of drinking during pregnancy are warning labels on alcoholic beverages and counseling during prenatal care. The NMIHS cannot provide any information on women's exposure to warning labels, since the label law did not go into effect until 1989; however, it does provide information on prenatal care and on the content of medical advice women received during prenatal care visits. Almost all the women in the NMIHS received prenatal care at some point in their pregnancies. Only 5.8 percent received no care or care beginning in the third trimester. Eighty percent of women made their first prenatal care visit in the first trimester, 14 percent in the second trimester, and 1.4 percent in the third trimester. The remaining 4.4 percent received no prenatal care.

Among all women who received prenatal care, 61 percent reported receiving advice to reduce drinking. However, a substantial proportion of women at risk of drinking during pregnancy did not receive any advice to modify their drinking behavior. In fact, 29 percent of women who reported drinking before pregnancy did not receive any advice on alcohol consumption during pregnancy, suggesting that despite almost universal levels of prenatal care among this population, not all women receive medical advice about drinking. As seen in Table A.4, overall, women were more likely to receive advice on proper nutrition and smoking than on drinking. Moreover, among the reasons given for reduced drinking during pregnancy, the advice of a doctor plays a relatively small role, with fewer than 20 percent reporting they reduced drinking at a doctor's urging (see Table 8 in Chapter 5).

As Table A.5 shows, disadvantaged women are more likely to receive late or no prenatal care. Black and other minority women, women with less than twelve years of

Table A.5. Odds of Receiving Prenatal Care and of
Receiving Advice to Reduce Drinking during Pregnancy,
Estimated Using the 1988 NMIHS

Characteristic (reference category)	Odds of Receiving Late or No Prenatal Care (N = 9,953)	Odds of Receiving Advice to Reduce Drinking (N = 9,953)
Race (white)		
Black	1.57**	0.86**
Other	1.59**	0.75**
Years of education (12)		
<12	1.42**	0.87**
13–16	0.72**	1.07
17+	0.43**	0.96
Age (20–29)		
<20	1.45**	0.89**
30–34	0.78*	1.01
35+	1.32**	0.75**
Parity (1)		
2	0.90	0.94
3+	1.29**	0.87
Unmarried (married)	2.20**	1.08
Income ($35,000+)		
<$10,000	2.38**	0.80**
$10,000–19,999	1.85**	0.83**
$20,000–34,999	1.67**	0.86**
Pre-pregnancy drinking level (low or none)		
Medium risk	1.53**	2.02**
High risk	3.26**	3.35***
Timing of prenatal care (1st trimester)		
2d trimester		0.78**
Late or none		0.08**

$*p < 0.05$
$**p < 0.01$

schooling, teenagers, women aged thirty-five and older, women with three or more children, unmarried women, and low-income women are all significantly more likely to begin prenatal care late in pregnancy or not at all. Drinking at high risk levels before pregnancy is also associated with a markedly higher odds of receiving late or no prenatal care.

Moreover, disadvantaged women are significantly less likely to receive advice to reduce drinking during pregnancy, even when controlling for pre-pregnancy drinking levels. Black women are 14 percent less likely than white women to receive advice, and women in other minorities are 25 percent less likely. Women who have not completed high school are 13 percent less likely than high school graduates to receive advice about drinking. There are no significant differences between women who have twelve years of schooling and women with college or graduate schooling. Both teenagers and women thirty-five years and older are less likely than women in their twenties to receive advice about drinking. There are no significant differences in the receipt of advice by parity or marital status. The odds of receiving advice increase with income. Moreover, the higher a woman's initial drinking level, the more likely she is to receive advice to reduce drinking. As might be expected, beginning prenatal care in the second trimester reduces the odds of receiving advice by 22 percent, and receiving late or no care reduces the odds almost 100 percent.

NOTES

Chapter 1. Conceiving Risk

1. *State of Wisconsin v Deborah J. Z.*, No. 96-2797-CR, Court of Appeals of Wisconsin, District Two (1999 Wisc App), LEXIS 581.

2. Zimmerman made repeated attempts over the next few years to regain custody of Megan, but she was unsuccessful. The girl was adopted before she turned three years old. In 1999, Mark Pinter and Colleen Zenk Pinter, both soap opera actors from Connecticut, bought the rights to Zimmerman's story. They hope to produce a television drama called "Held Accountable: The Deborah Zimmerman Story." Tom Kertscher, "Acting couple obtain rights to story of drinking mom," *Milwaukee Journal Sentinel*, 15 Aug. 1999.

3. Meg Jones, "Prosecutor wants mother on trial," *Milwaukee Journal Sentinel*, 6 Sept. 1996.

4. Zimmerman's case was appealed all the way to the Wisconsin Supreme Court, but that court was unable to reach a decision on the case and therefore remanded it to the Wisconsin Court of Appeals.

5. Kathleen Stratton, Cynthia Howe, and Frederick Battaglia, eds., *Fetal Alcohol Syndrome: Diagnosis, Epidemiology, Prevention, and Treatment* (Washington, D.C.: National Academy Press, 1996).

6. Ibid.

7. Gilberto F. Chavez, José F. Cordero, and José E. Becerra, "Leading major congenital malformations among minority groups in the United States, 1981–1986," *Morbidity and Mortality Weekly Report* 37, no. SS-3 (1 July 1988): 17–24.

8. Centers for Disease Control, "Fetal alcohol syndrome — United States, 1979–1992," *Morbidity and Mortality Weekly Report* 42, no. 17 (7 May 1993): 339–41; Jon M. Aase, "Clinical recognition of FAS: difficulties of detection and diagnosis," *Alcohol Health and Research World* 18, no. 1 (1994): 5–9.

9. Aase.

10. Stratton, Howe, and Battaglia, 76–77.

11. Ibid., 182.

12. Ernest L. Abel, "An update on the incidence of FAS: FAS is not an equal opportunity birth defect," *Neurotoxicology and Teratology* 17, no. 4 (1995): 437–43.

13. Ernest L. Abel and John H. Hannigan, "Maternal risk factors in fetal alcohol syndrome: provocative and permissive influences," *Neurotoxicology and Teratology* 17, no. 4 (1995): 445–62.

14. Katherine K. Christoffel and Ira Salafsky, "Fetal alcohol syndrome in dizygotic twins," *Journal of Pediatrics* 87, no. 6 (1975): 963–67.

15. Ernest L. Abel, *Fetal Alcohol Abuse Syndrome* (New York: Plenum Press, 1998).

16. Ernest L. Abel, "Fetal alcohol syndrome in families," *Neurotoxicology and Teratology* 10, no. 1 (1988): 1–2.

17. Abel, "Update on incidence of FAS," 437–43.

18. Centers for Disease Control, "Sociodemographic and behavioral characteristics associated with alcohol consumption during pregnancy — United States, 1988," *Morbidity and Mortality Weekly Report* 44, no. 13 (7 Apr. 1995): 1118–20.

19. Centers for Disease Control, "Alcohol consumption among pregnant and childbearing-aged women — United States, 1991 and 1995," *Morbidity and Mortality Weekly Report* 46, no. 16 (25 Apr. 1997): 346–50.

20. Renée C. Fox, "The Medicalization and Demedicalization of American Society," in *Essays in Medical Sociology: Journeys into the Field* (New Brunswick, N.J.: Transaction Press, 1988).

21. Eliot Freidson, *Profession of Medicine: A Study of the Sociology of Applied Knowledge* (New York: Dodd Mead, 1970), 245.

22. Allan M. Brandt and Paul Rozin, "Introduction," in *Morality and Health*, ed. Allan M. Brandt and Paul Rozin (New York: Routledge, 1997), 5.

23. Talcott Parsons, "Social Structure and Dynamic Process: The Case of Modern Medical Practice," in *The Social System* (Glencoe, Ill.: Free Press, 1951), 429.

24. Charles E. Rosenberg, "Banishing risk: or the more things change the more they stay the same," *Perspectives in Biology and Medicine* 39, no. 1 (1995): 28.

25. Brandt and Rozin, 4.

26. Kristin Luker, *Abortion and the Politics of Motherhood* (Berkeley: University of California Press, 1984); Emily Martin, *The Woman in the Body* (Boston: Beacon Press, 1987).

27. George J. Annas, "Pregnant women as fetal containers," *Hastings Center Report* 16, no. 6 (1986): 13.

28. Michel Foucault, *The Birth of the Clinic: An Archeology of Medical Perception* (New York: Pantheon Books, 1973); Catherine Kohler Riessman and Constance A. Nathanson, "The Management of Reproduction: Social Construction of Risk and Responsibility," in *Applications of Social Science to Clinical Medicine and Health Policy*, ed. Linda H. Aiken and David Mechanic (New Brunswick, N.J.: Rutgers University Press, 1986).

29. William Ray Arney, *Power and the Profession of Obstetrics* (Chicago: University of Chicago Press, 1982), 91.

30. Ana V. Diez-Roux, "Bringing context back into epidemiology: variables and fallacies in multilevel analysis," *American Journal of Public Health* 88, no. 2 (1998): 216–22; Irving K. Zola, "Medicine as an institution of social control," *Sociological Review* 20 (1972): 487–504.

31. Zola.

32. Elliot G. Mishler, "The Social Construction of Illness," in *Social Contexts of Health, Illness, and Patient Care,* ed. Elliot G. Mishler (Cambridge: Cambridge University Press, 1981), 162.

33. Martin; Barbara Katz Rothman, *The Tentative Pregnancy: Prenatal Diagnosis and the Future of Motherhood* (New York: Penguin Books, 1986); Robert H. Blank, *Regulating Reproduction* (New York: Columbia University Press, 1990).

34. Blank.

35. Viviana A. Zelizer, *Pricing the Priceless Child: The Changing Social Value of Children* (Princeton: Princeton University Press, 1985).

36. Arney; Cynthia R. Daniels, *At Women's Expense: State Power and the Politics of Fetal Rights* (Cambridge: Harvard University Press, 1993); Steven Maynard-Moody, *The Dilemma of the Fetus: Fetal Research, Medical Progress, and Moral Politics* (New York: St. Martin's Press, 1995); Monica Casper, *The Making of the Unborn Patient: A Social Anatomy of Fetal Surgery* (New Brunswick, N.J.: Rutgers University Press, 1998).

37. Plato, *Laws,* trans. Thomas L. Pangle (New York: Basic Books, 1980), 6:775.

38. C. W. Saleeby, "Professor Karl Pearson on alcoholism and offspring," *British Journal of Inebriety* 8, no. 2 (Oct. 1910): 53–66.

39. Charles E. Rosenberg, "The Bitter Fruit: Heredity, Disease and Social Thought," in *No Other Gods: On Science and American Social Thought* (Baltimore: Johns Hopkins University Press, 1976).

40. Daniels; Luker.

41. Carol Brooks Gardner, "The Social Construction of Pregnancy and Fetal Development: Notes on a Nineteenth-Century Rhetoric of Endangerment," in *Constructing the Social,* ed. Theodore R. Sarbin and John I. Kitsuse (London: Sage, 1994).

42. Barbara Duden, *Disembodying Women: Perspectives on Pregnancy and the Unborn* (Cambridge: Harvard University Press, 1993); Gardner.

43. Parsons, 448.

44. Jane Balin, "The sacred dimensions of pregnancy and birth," *Qualitative Sociology* 11, no. 4 (1988): 277.

45. Arney, 94.

46. Daniels.

47. Charles E. Rosenberg and Carroll Smith-Rosenberg, "The Female Animal: Medical and Biological Views of Women," in C. E. Rosenberg, *No Other Gods.*

48. C. Wright Mills, *The Sociological Imagination* (New York: Oxford University Press, 1959).

Chapter 2. The "Question of Alcohol and Offspring" in the Nineteenth Century

1. Plato, *Laws*, trans. Thomas L. Pangle (New York: Basic Books, 1980), 6:775.

2. Quoted in Ernest L. Abel, *Fetal Alcohol Syndrome and Fetal Alcohol Effects* (New York: Plenum Press, 1984).

3. Soranus, *Gynecology*, trans. Oswei Temkin (Baltimore: Johns Hopkins University Press, 1956; reprint, 1991).

4. Ambroise Paré, *Monsters and Marvels*, trans. Janis L. Pallister (Chicago: University of Chicago Press, 1982), 38.

5. Even today such a birthmark is referred to as a strawberry mark.

6. Paré, 3.

7. The notion that drunkenness at conception adversely affected the child is a remarkably persistent one. Even the current entry on fetal alcohol syndrome in *Encyclopaedia Britannica* (1998) reflects this belief, defining FAS as "various congenital abnormalities in the newborn infant that are caused by the mother's ingestion of significant amounts of alcohol *around the time of conception* or during pregnancy" (emphasis added).

8. Rebecca H. Warner and Henry L. Rosett, "The effects of drinking on offspring: an historical survey of the American and British literature," *Journal of Studies on Alcohol* 36 (1975): 1397.

9. Ibid.

10. Ibid.

11. Quoted in Abel.

12. Henry Fielding, *An Enquiry into the Causes of the Late Increase of Robbers, etc. with Some Proposals for Remedying This Growing Evil* (London: A. Millar, 1751).

13. Warner and Rosett, 1397.

14. Edward Barry, *Essays, on the Following Subjects: Celibacy, Wedlock, Seduction, Pride, Duelling, Self-Murder, Lying, Detraction, Avarice, Justice, Generosity, Temperance, Excess, Death* (Reading, U.K.: Smart and Cowslade, 1806).

15. Charles E. Rosenberg, "The Bitter Fruit: Heredity, Disease and Social Thought," in *No Other Gods: On Science and American Social Thought* (Baltimore: Johns Hopkins University Press, 1976); Edwin H. Ackerknecht, "Diathesis: the word and concept in medical history," *Bulletin of the History of Medicine* 56 (1982): 317.

16. Ackerknecht, 317.

17. Robert C. Olby, "Constitutional and Hereditary Disorders," in *Companion Encyclopedia of the History of Medicine*, vol. 1, ed. William F. Bynum and Roy Porter (London: Routledge, 1993).

18. James Whitehead, *On the Transmission, from Parent to Offspring, of Some Forms of Disease, and of Morbid Taints and Tendencies*, 2d ed. (London: John Churchill, 1857).

19. Nathan Allen, "Effects of alcohol on the offspring," *Journal of Psychological Medical and Mental Pathology* (London), n.s., 3 (1877): 210.

20. Olby.

21. Rosenberg, "Bitter Fruit."

22. S. A. K. Strahan, *Marriage and Disease: A Study of Heredity and the More Important Family Degenerations* (London: K. Paul, Trench, Trübner), excerpted in *Quarterly Journal of Inebriety* 15, no. 1 (1893): 52–59.

23. *Report from the Select Committee on Inquiry into Drunkenness* (London: House of Commons, 1834), v.

24. Ibid., 219.

25. Samuel G. Howe, *On the Causes of Idiocy* (New York: Arno Press, 1848; reprint, 1972), 29.

26. General Assembly of Connecticut, "Causes of Idiocy" (reported by the Commissioners on Idiocy; Hartford: General Assembly of Connecticut, 1856), 9.

27. Allen.

28. However, studies of defective or captive populations such as these continued through the first decade of the twentieth century. See, e.g., T. Alexander MacNicholl, "A study of the effects of alcohol on schoolchildren" *Quarterly Journal of Inebriety* 27, no. 2 (1905): 113–17; Louise G. Robinovitch, "The relation of alcoholism in the parent to idiocy and imbecility of the offspring. Chapter I. Hereditary degeneracy," *Journal of Mental Pathology* 1, no. 1 (1901): 14–24; Louise G. Robinovitch, "The relation of alcoholism in the parent to idiocy and imbecility of the offspring. Chapter II. The relation of alcoholism in the parent to idiocy and imbecility of the offspring," *Journal of Mental Pathology* 1, no. 1 (1901): 86–96.

29. Dom. Bezzola, "A statistical investigation into the role of alcohol in the origin of innate imbecility," *Quarterly Journal of Inebriety* 23 (1901): 346–54.

30. Ibid.

31. Ibid.

32. Strahan.

33. "Le Progrès Médicale," *Quarterly Journal of Inebriety* 15, no. 1 (1893): 50.

34. Thomas Davison Crothers, "Inebriety traced to the intoxication of parents at the time of conception," *Medical and Surgical Reporter* 56 (1887): 549–51.

35. Strahan, 52.

36. Howe, 28.

37. *Inquiry into Drunkenness*, 219.

38. Henry H. Goddard, "Alcoholism and feeble-mindedness," *Interstate Medical Journal* 23 (1916): 445.

39. *Kallikak* was a pseudonym invented by Goddard, from the Greek words for "good" and "bad."

40. J. M. French, "A study of the hereditary effects of alcohol," *Medicine* (Detroit) 8 (1902): 22.

41. Alfred Gordon, "The influence of alcohol on the progeniture," *Interstate Medical Journal* 23, no. 6 (1916): 432.

42. Rosenberg, "Bitter Fruit," 43.

43. Ibid.; Jean-Charles Sournia, *A History of Alcoholism*, trans. Nick Hindley and Gareth Stanton (Cambridge, Mass.: Basil Blackwell, 1990).

44. Maurice Legrain, *Quarterly Journal of Inebriety* 15, no. 2 (1900): 162.

45. John T. Irion, "The effects of alcohol on offspring," *Mississippi Valley Medical Monthly* 4 (1884): 384.

46. Sournia, 99.

47. Ibid. Ironically, about the time that *degeneration*, the term and the concept, fell out of medical usage it seems to have been picked up by social scientists to explain persistent poverty in families across generations. In other words, it became an element in arguments about the culture of poverty.

48. Olby.

49. Ibid.

50. Sournia.

51. *Inquiry into Drunkenness*, 220.

52. Quoted in Warner and Rosett, 1402.

53. John Ellis, "Deterioration of the Puritan Stock and Its Causes" (pamphlet published by the author, New York, 1884), 30.

54. Allen, 209.

55. Editorial, *Quarterly Journal of Inebriety* 15, no. 3 (1893): 297.

56. R. M. Wigginton, *Quarterly Journal of Inebriety* 15, no. 4 (1893): 337.

57. Allen, 214.

58. Howe, 4.

59. William C. Sullivan, "A note on the influence of maternal inebriety on the offspring," *Journal of Mental Science* 45 (1899): 489–503.

60. Leonard Darwin, "Alcoholism and eugenics," *British Journal of Inebriety* 13, no. 2 (1915): 55–86.

61. Strahan, 56.

62. Rosenberg, "Bitter Fruit," 43.

63. Thomas Davison Crothers, "Alcoholic heredity in diseases of children," *Journal of the American Medical Association* 15 (1890): 533.

64. Olby, 421.

65. Harry S. Warner, *Social Welfare and the Liquor Problem* (Chicago: Intercollegiate Prohibition Association, 1913), 90.

66. In this period, the terms *germ cells* and *germ plasm* were used to refer to the egg and sperm, the gametes, or germinative cells.

67. Leslie E. Keeley, *The Non-Heredity of Inebriety* (Chicago: Scott, Foresman, 1896), 68.

68. Crothers, "Degeneration common to alcoholism," 109.

69. D. H. Kress, "The use of narcotics as related to the declining birth rate and race suicide," *Journal of Inebriety* 33 (1911): 71.

70. August Forel, "The effect of alcoholic intoxication upon the human brain and its relation to theories of heredity and evolution," *Quarterly Journal of Inebriety* 15, no. 3 (1893): 203–21.

71. Thomas Trotter, *An Essay, Medical, Philosophical, and Chemical, on Drunkenness, and Its Effects on the Human Body* (Boston: Bradford and Read, 1813).

72. Forel.

73. Ibid., 215.

74. Quoted in Daniel J. Kevles, *In the Name of Eugenics: Genetics and the Uses of Human Heredity* (New York: Knopf, 1985).

75. Ibid.

76. Ethel M. Elderton and Karl Pearson, *A First Study of the Influence of Parental Alcoholism on the Physique and Ability of the Offspring* (London: Dulau, 1910), 1.

77. Karl Pearson and Ethel M. Elderton, *A Second Study of the Influence of Parental Alcoholism on the Physique and Ability of the Offspring, Being a Reply to Certain Medical Critics of the First Memoir and an Examination of the Rebutting Evidence Cited by Them* (London: Dulau, 1910).

78. Elderton and Pearson.

79. Pearson and Elderton.

80. Elderton and Pearson.

81. C. W. Saleeby, "Professor Karl Pearson on alcoholism and offspring," *British Journal of Inebriety* 8, no. 2 (1910): 53–66.

82. T. A. Chapman, "Alcoholism and degeneracy" (correspondence), *British Medical Journal* 2 (31 Dec. 1910): 2049.

83. Mary D. Sturge and Victor Horsley, "Alcoholism and degeneration" (correspondence), *British Medical Journal* 2 (1910): 1656.

84. Pearson and Elderton, 8.

85. T. Morton, "The problem of heredity in reference to inebriety," *Quarterly Journal of Inebriety* 17, no. 1 (1895): 7.

86. C. W. Saleeby, "Alcoholism and eugenics," *British Journal of Inebriety* 7, no. 1 (1909): 13.

87. Gordon, 436

88. William McAdam Eccles, "Alcohol as a factor in the causation of deterioration in the individual and the race," *British Journal of Inebriety* 2, no. 4 (1905): 154.

89. Clarence True Wilson and Deets Pickett, *The Case for Prohibition: Its Past, Present Accomplishments, and Future in America* (New York: Funk and Wagnalls, 1923), 79–80.

90. Saleeby, "Alcoholism and eugenics," 7.

91. G. Archdall Reid, "Human evolution: with especial reference to alcohol," *British Medical Journal* 2 (3 Oct. 1903): 818–20; with comments by Charles Mercier, David Yellowlees, T. N. Kelynack, F. Edridge-Green, A. T. Schofield, J. J. Ridge, J. F. Elashman, W. Ford Robertson, W. Lloyd Andriezen, Heywood Smith, Fletcher Beach, Wilfred Harris, James Stewart, and Robert Jones.

92. Ibid.

93. G. Archdall Reid, *Alcoholism: A Study in Heredity* (New York: William Wood, 1902).

94. Darwin, 69.

95. Reid, *Alcoholism*, 31.

96. Reid, "Human evolution," 819–200.

97. Reid, *Alcoholism*, 167.

98. Ibid., 171.

99. Albert Edward Wiggam, *The Next Age of Man* (New York: Blue Ribbon Books, 1927), 46.

100. Hobart A. Hare, "A lecture on alcohol and its use in medicine," *Therapeutic Gazette* 49 (1925): 20.

101. Bénédict Auguste Morel, *Traité des dégénérescences physiques, intellectuelles, et morales de l'éspèce humaine et des causes qui produisent ces variétés maladives* (Paris: Bailliére, 1857).

102. Wiggam, 47.

103. Charles Loomis Dana, "Nervous and mental diseases and the Volstead Law," *North American Review* 221 (1925): 617.

104. Keeley, 154.

105. William B. Drummond, "Alcoholism in relation to heredity" (correspondence), *British Medical Journal* 2 (21 Oct. 1899): 1137.

106. Casper L. Redfield, "Kallikaks and Jukes," *Detroit Medical Journal*, 19 (1918): 405.

107. J. W. Ballantyne, "Alcohol as a beverage in its relation to infantile mortality," *British Medical Journal* 2 (12 Aug. 1922): 254.

108. Quoted in Society for the Study of Inebriety, "Is inebriety hereditary?" *Interstate Medical Journal* 6 (1899): 388.

109. "Alcohol and the germ plasm" (editorial), *Journal of the American Medical Association* 82, no. 18 (1924): 1444.

110. Quoted in Bartlett C. Jones, "Prohibition and eugenics: 1920–1933," *Journal of History of Medicine and Allied Sciences* 18 (1963): 159.

111. Charles R. Stockard, "Alcohol as a selective agent in the improvement of racial stock," *British Medical Journal* 2 (12 Aug. 1922): 258.

112. Henry H. Goddard, *Feeble-mindedness: Its Causes and Consequences* (New York: Macmillan, 1923), 491.

113. Hare, 20.

114. Harry Campbell, "Reply to J. W. Ballantyne 'Alcohol and antenatal child welfare'," *British Journal of Inebriety* 14, no. 3 (1917): 125.

115. Campbell, 125.

116. Gerrit Pieter Frets, *Alcohol and the Other Germ Poisons* (The Hague: Martinus Nijhoff, 1931).

117. Stockard, "Alcohol as a selective agent," 255.

118. Ibid., 256.

119. Charles R. Stockard, "Alcohol a factor in eliminating racial degeneracy," *American Journal of Medical Science* 167, no. 4 (1924): 469–77.

120. Raymond Pearl, "The experimental modification of germ cells. III. The effect

of parental alcoholism, and certain other drug intoxications, upon the progeny," *Journal of Experimental Zoology* 22, no. 2 (1917): 241–99.

121. Stockard, "Alcohol a factor in eliminating racial degeneracy," 476–77.

122. Stockard, "Alcohol as a selective agent," 258.

123. Mary Scharlieb, "Alcohol and the child-bearing woman," *British Journal of Inebriety* 11, no. 2 (1913): 62.

124. Saleeby, "Alcoholism and eugenics," 16.

125. J. Matthews Duncan, "On alcoholism in gynaecology and obstetrics," *Transactions of the Edinburgh Obstetrical Society* 13 (1888): 105–20.

126. Darwin, 69.

127. Eccles, 153.

128. "Some causes of congenital deformities" (editorial, Clinical Notes and Comments), *Quarterly Journal of Inebriety* 21 (1899): 203.

129. J. W. Ballantyne, "Alcoholism and antenatal hygiene," *British Journal of Inebriety* 13, no. 2 (1915): 88.

130. Sullivan, "Influence of maternal inebriety," 489–503.

131. Ibid.

132. Ibid.

133. Ballantyne, "Alcoholism and antenatal hygiene," 88.

134. Duncan, "Alcoholism in gynaecology and obstetrics," 116.

135. B. C. Keister, "The social glass, a menace to civilization," *Journal of Inebriety* 33 (1911): 117–24.

136. Lawrence Irwell, "Influence of parental alcoholism upon the human family," *Medical Times* (New York) 41, no. 4 (1913): 114.

137. Ibid.

138. Thomas Davison Crothers, "A new view of alcoholic degeneration," *Medical Times* (New York) 44, no. 7 (1916): 219.

139. Kathleen Stratton, Cynthia Howe, and Frederick Battaglia, eds., *Fetal Alcohol Syndrome: Diagnosis, Epidemiology, Prevention, and Treatment* (Washington, D.C.: National Academy Press, 1996).

140. A. Louise McIlroy, "Alcohol as a beverage in its relation to infantile mortality" (reply to J. W. Ballantyne), *British Medical Journal* 2 (12 Aug. 1922): 259.

141. C. W. Stewart, "Alcohol as a factor in rendering mothers incapable of nursing their children" *Quarterly Journal of Inebriety* 27, no. 1 (1905): 36.

142. Keister, 124.

143. Strahan, 59.

144. Sullivan, "Influence of maternal inebriety," 494.

145. Casper L. Redfield, "Alcoholism and heredity," *Medical Review of Reviews* 26 (1920): 308.

146. T. Morton, "The problem of heredity in reference to inebriety," *Quarterly Journal of Inebriety* 17, no. 1 (1895): 8.

147. Saleeby, "Alcoholism and eugenics," 7–20.

148. C. C. Wholey, "Alcohol and heredity," *West Virginia Medical Journal* 7 (Feb. 1913): 262.

149. Ballantyne, "Alcohol in relation to infantile mortality," 259.

150. A. Louise McIlroy, "The influence of alcohol and alcoholism upon antenatal and infant life," *British Journal of Inebriety* 21, no. 2 (1923): 42.

151. Keeley, 217.

152. Irwell, 114.

153. Keeley, 334.

154. Scharlieb.

155. Keister, 124.

156. John Haddon, "On intemperance in women, with special reference to its effects on the reproductive system," *British Medical Journal* 1 (17 June 1876): 748.

157. McIlroy, "Influence of alcohol and alcoholism," 42. Note that the concern about female drinking in this period is an inversion of the anxiety expressed during the Gin Epidemic. Then, commentators feared drinking among the lower classes; here, it is drinking among the "wealthier classes" that causes alarm.

158. Charles E. Rosenberg and Carroll Smith-Rosenberg, "The Female Animal: Medical and Biological Views of Women," in Charles E. Rosenberg, *No Other Gods: On Science and American Social Thought* (Baltimore: Johns Hopkins University Press, 1976).

159. Scharlieb, 62.

160. William C. Sullivan, "The children of the female inebriate," *Quarterly Journal of Inebriety* 22, no. 2 (1900): 137.

161. Rosenberg and Smith-Rosenberg, 61.

162. Dr. Ayers, "Wine drinking during pregnancy" (editorial), *Quarterly Journal of Inebriety* 23 (1901): 76.

163. Sullivan, "Children of female inebriate," 138.

164. Irwell, 114.

165. Karl Mannheim, *Ideology and Utopia* (New York: Harvest Books, 1936), 265.

166. Warner, 91.

167. Ibid., 91.

168. Morton, "Heredity in reference to inebriety," 1.

169. Rosenberg, "Bitter Fruit."

170. Ballantyne, "Alcohol and antenatal child welfare," 106.

171. Alfred Gordon, "Parental alcoholism as a factor in the mental deficiency of children: a statistical study of 117 families," *Journal of Inebriety* 35, no. 2 (1913): 64.

172. Rosenberg, "Bitter Fruit," 35.

173. Ibid.

174. Warner and Rosett, 1397; Abel.

175. Otto L. Mohr, *Heredity and Disease* (New York: W. W. Norton, 1934), 195.

176. Howard H. Haggard and E. M. Jellinek, *Alcohol Explored* (New York: Doubleday, Doran, 1942): 207.

Chapter 3. Diagnosing Moral Disorder

1. National Institute on Alcohol Abuse and Alcoholism (NIAAA), *U.S. Apparent Consumption of Alcoholic Beverages*, vol. 1, 3d ed. (Bethesda, Md.: NIAAA, 1997).

2. Catherine Gilbert Murdock, *Domesticating Drink: Women, Men, and Alcohol in America, 1870–1940* (Baltimore: Johns Hopkins University Press, 1998).

3. Francis Sill Wickware, "Liquor: current studies in medicine and psychiatry are bringing enlightenment to the 30,000-year-old problem of drinking," *Life*, 27 May 1946.

4. Ron Roizen, "How Does the Nation's 'Alcohol Problem' Change from Era to Era? Stalking the Social Logic of Problem-Definition Transformations since Repeal" (paper presented at the Francis Clark Wood Institute for the History of Medicine of the College of Physicians of Philadelphia, 9–11 May 1997).

5. Otto L. Mohr, *Heredity and Disease* (New York: W. W. Norton, 1934), 195.

6. "Effect of single large alcohol intake on fetus" (editorial, Queries and Minor Notes), *Journal of the American Medical Association* 120, no. 1 (1942): 88.

7. Roizen.

8. Howard W. Haggard and E. M. Jellinek, *Alcohol Explored* (New York: Doubleday, Doran, 1942), 207.

9. Wickware.

10. Clifford B. Lull and Robert A. Kimbrough, eds., *Clinical Obstetrics* (Philadelphia: J. B. Lippincott, 1953), 160–61.

11. Charles E. McLennan and Eugene C. Sandberg, *Synopsis of Obstetrics*, 8th ed. (St. Louis: C. V. Mosby, 1970).

12. Mae M. Bookmiller and George L. Bowen, *Textbook of Obstetrics and Obstetric Nursing*, 4th ed. (Philadelphia: W. B. Saunders, 1963), 136.

13. Jean Towler and Roy Butler-Manuel, *Modern Obstetrics for Student Midwives* (Chicago: Year Book Medical Publishers, 1973), 145.

14. Kenneth R. Niswander, ed., *Obstetrics: Essentials of Clinical Practice* (Boston: Little, Brown, 1976).

15. In this section I occasionally draw on the interviews with doctors reported in Chapter 4. See that chapter for a full description (doctors are identified by ID number within brackets, as listed in Table 4).

16. See, e.g., Fritz Fuchs, Anna-Riitta Fuchs, Vincente F. Poblete, and Abraham Risk, "Effect of alcohol on threatened premature labor," *American Journal of Obstetrics and Gynecology* 99 no. 5 (1967): 627–37.

17. Wolfgang Saxon, "Dr. Fritz Fuchs, 76, who advanced obstetrics," *New York Times*, 4 Mar. 1995, sect. 1, 26.

18. Rebecca H. Warner and Henry L. Rosett, "The effects of drinking on offspring: an historical survey of the American and British literature," *Journal of Studies on Alcohol* 36 (1975): 1397.

19. According to *Taber's Cyclopedic Medical Dictionary* (16th ed.), a syndrome is

"a group of symptoms and signs of disordered function related to one another by means of some anatomic, physiologic, or biochemical peculiarity."

20. In order to scrutinize the initial process of creating and establishing FAS as a diagnosis, I searched the medical literature for the period 1966–84 using the Medline database, which includes articles dating from 1966 to the present. Since *fetal alcohol syndrome* is not available as a keyword for searching articles published before 1976, I also used the keyword combinations *alcohol drinking* and *pregnancy, ethyl alcohol* and *pregnancy*, and *ethyl alcohol* and *abnormalities* for the entire period. The scope of the review was limited to English-language articles concerning human subjects. (There is also an abundant literature on the effects of alcohol during pregnancy in animal models.) I reviewed only a limited number of articles on the associations between alcohol use and spontaneous abortion and between alcohol and low birthweight. For the period 1966–72, before publication of the first articles on FAS, I found more than a hundred items (including articles, letters to the editor, and brief reports) on alcohol and pregnancy, the majority of these concerning the intravenous administration of ethanol to arrest preterm labor. I concentrate here on the literature from 1973 to 1984; however, I occasionally refer to more recent medical literature on FAS.

21. Howard S. Becker, *Outsiders: Studies in the Sociology of Deviance* (New York: Free Press, 1963).

22. Eliot Freidson, *Profession of Medicine: A Study of the Sociology of Applied Knowledge* (New York: Dodd Mead, 1970).

23. Stephen Hilgartner and Charles L. Bosk, "The rise and fall of social problems: a public arenas model," *American Journal of Sociology* 94, no. 1 (1988): 53–78.

24. Food and Drug Administration, "Surgeon general's advisory on alcohol and pregnancy," *FDA Drug Bulletin* 11, no. 2 (1981): 9–10. The warning is "The Surgeon General advises women who are pregnant (or considering pregnancy) not to drink alcoholic beverages and to be aware of the alcoholic content of foods and drugs."

25. David W. Smith, *Recognizable Patterns of Human Malformation: Genetic, Embryonic, and Clinical Aspects* (Philadelphia: W. B. Saunders, 1970); David W. Smith, *Recognizable Patterns of Human Malformation: Genetic, Embryonic, and Clinical Aspects*, 2d ed. (Philadelphia: W. B. Saunders, 1976).

26. Centers for Disease Control, "Temporal trends in the incidence of birth defects — United States," *Morbidity and Mortality Weekly Report* 46, no. 49 (12 Dec. 1997): 1171–76.

27. Kenneth L. Jones, David W. Smith, Christy N. Ulleland, and Ann P. Streissguth, "Pattern of malformation in offspring of chronic alcoholic mothers," *Lancet* 1, no. 7815 (9 June 1973): 1267–71.

28. Ibid., 1267.

29. Ibid., 1271.

30. Kenneth Lyon Jones and David W. Smith, "Recognition of the fetal alcohol syndrome in early infancy," *Lancet* 2, no. 7836 (3 Nov. 1973): 999–1001.

31. Kenneth Lyon Jones, David W. Smith, Ann P. Streissguth, and Ntinos C. Myr-

ianthopoulos, "Outcome in offspring of chronic alcoholic women," *Lancet* 1, no. 7866 (1 June 1974): 1076–78.

32. This article (P. Lemoine, H. Harousseau, J. P. Borteryu, and J. C. Menuet, "Les enfants de parents alcooliques: anomalies observées à propos de 127 cas," *Ouest Médical* 21 [1968]: 476–82) was not recognized in the English-language medical literature until *after* 1973; however, it is often reconstructed as "the first report of FAS."

33. See, e.g., P. E. Ferrier, I. Nicod, and S. Ferrier, "Fetal alcohol syndrome" (letter to the editor), *Lancet* 2, no. 7844 (1973): 1496; Katherine K. Christoffel and Ira Salafsky, "Fetal alcohol syndrome in dizygotic twins," *Journal of Pediatrics* 87, no. 6 (1975): 963–67; Margaret S. Tenbrinck and Sandra Y. Buchin, "Fetal alcohol syndrome: report of a case," *Journal of the American Medical Association* 232, no. 11 (1975): 1144–47; John J. Mulvihill, John T. Klimas, Dennis C. Stokes, and Herman M. Risemberg, "Fetal alcohol syndrome: seven new cases," *American Journal of Obstetrics and Gynecology* 125, no. 7 (1976): 937–41; James W. Hanson, Kenneth L. Jones, and David W. Smith, "Fetal alcohol syndrome: experience with 41 patients," *Journal of the American Medical Association* 235, no. 14 (1976): 1458–60; Elizabeth K. Turner, "Fetal alcohol syndrome" (letter to the editor), *Medical Journal of Australia* 66, no. 5 (1979): 178.

34. Ferrier et al., 1496.

35. Jones and Smith, "Recognition of the fetal alcohol syndrome."

36. Mulvihill et al.

37. Hanson et al.

38. David W. Smith, "The fetal alcohol syndrome," *Hospital Practice* 14, no. 10 (1979): 121–28.

39. J. H. Renwick and R. L. Asker, "Ethanol-sensitive times for the human conceptus," *Early Human Development* 8, no. 2 (1983): 99.

40. Herbert Blumer, "Social problems as collective behavior," *Social Problems* 18 (1971): 298–306.

41. Ibid.

42. R. Sturdevant, "Offspring of chronic alcoholic women" (letter to the editor), *Lancet* 2, no. 7876 (1974): 349.

43. Charles F. Johnson, "Does maternal alcoholism affect offspring?" *Clinical Pediatrics* 13, no. 8 (1974): 633–34.

44. Vipul N. Mankad and Rajendra M. Choksi, "The fetal alcohol syndrome" (letter to the editor), *Journal of the American Medical Association* 236, no. 10 (1976): 1114.

45. R. C. Sneed, "The fetal alcohol syndrome: is alcohol, lead, or something else the culprit?" (letter to the editor), *Journal of Pediatrics* 90, no. 2 (1977): 324; Jack H. Mendelson, "The fetal alcohol syndrome" (letter to the editor), *New England Journal of Medicine* 299, no. 10 (1978): 556; J. Wilson, "The fetal alcohol syndrome," *Public Health* 95, no. 3 (1981): 129–32; Arthur J. Amman, Diane W. Wara, Morton J. Cowan, Douglas J. Barett, and E. Richard Stiehm, "The DiGeorge syndrome and the fetal alcohol syndrome," *American Journal of Diseases of Children* 136, no. 10 (1982): 906–8; Ayhan O. Çavdar, "DiGeorge's syndrome and fetal alcohol syndrome" (letter to the ed-

itor), *American Journal of Diseases of Children* 137, no. 8 (1983): 806–7; Ayhan O. Çavdar, "Fetal alcohol syndrome, malignancies, and zinc deficiency" (letter to the editor), *Journal of Pediatrics* 105, no. 2 (1984): 335.

46. Frank Majewski, "Alcohol embryopathy: some facts and speculations about pathogenesis," *Neurobehavioral Toxicology and Teratology* 3, no. 2 (1981): 129.

47. P. V. Véghelyi and Magda Osztovics, "The fetal alcohol syndrome: the intracombigenic effect of acetaldehyde," *Experientia* 34, no. 2 (1978): 195–96; P. M. Dunn, S. Stewart-Brown, and R. Peel, "Metronidazole and the fetal alcohol syndrome" (letter to the editor), *Lancet* 2, no. 8134 (1979): 144; Ernest L. Abel, "Consumption of alcohol during pregnancy: a review of effects on growth and development of offspring," *Human Biology* 54, no. 3 (1982): 421–53.

48. This hypothesis has recently been reintroduced by molecular biologists who suspect that a deficiency in the enzyme alcohol dehydrogenase, which breaks down ethanol, may leave some fetuses at increased risk of developing birth defects.

49. Joseph Gusfield, *The Culture of Public Problems: Drinking-Driving and the Symbolic Order* (Chicago: University of Chicago Press, 1981), 72.

50. Ernest L. Abel, *Fetal Alcohol Abuse Syndrome* (New York: Plenum Press, 1998).

51. Nesrin Bingol, Carlotta Schuster, Magdalena Fuchs, Silvia Iosub, Gudrun Turner, Richard K. Stone, and Donald S. Gromisch, "The influence of socioeconomic factors on the occurrence of fetal alcohol syndrome," *Advances in Alcohol and Substance Abuse* 6, no. 4 (1987): 105–18.

52. Ernest L. Abel and John H. Hannigan, "Maternal risk factors in fetal alcohol syndrome: provocative and permissive influences," *Neurotoxicology and Teratology* 17 no. 4 (1995): 445–62.

53. Dunn et al., 144.

54. Gerald F. Chernoff, "Introduction: A Teratologist's View of the Fetal Alcohol Syndrome," in *Currents in Alcoholism*, vol. 7, *Recent Advances in Research and Treatment*, ed. M. Galanter (New York: Grune and Stratton, 1979), 7.

55. Henry F. Krous, "Fetal alcohol syndrome: a dilemma of maternal alcoholism," *Pathology Annual* 16, no. 1 (1981): 307.

56. Gusfield, *Culture of Public Problems*, 43.

57. Jack P. Lipton and Alan M. Hershaft, "On the widespread acceptance of dubious medical findings," *Journal of Health and Social Behavior* 26 (1985): 336–51.

58. Gideon Koren, Karen Graham, Heather Shear, and Tom Einarson, "Bias against the null hypothesis: the reproductive hazards of cocaine," *Lancet* 2, no. 8677 (16 Dec. 1989): 1440–42.

59. Joel Best, *Threatened Children: Rhetoric and Concern about Child-Victims* (Chicago: University of Chicago Press, 1990).

60. Ibid., 65.

61. Sterling K. Clarren and David W. Smith, "The fetal alcohol syndrome" (reply to letter to the editor), *New England Journal of Medicine* 298, no. 19 (1978): 1063–67.

62. Elizabeth Rasche González, "New ophthalmic findings in fetal alcohol syndrome," *Journal of the American Medical Association* 245, no. 2 (1981): 108.

63. Ibid.

64. Qutub H. Qazi, Akiko Masakawa, Barbara McGann, and James Woods, "Dermatoglyphic abnormalities in the fetal alcohol syndrome," *Teratology* 21 (1980): 157. Abnormal dermatoglyphics are also characteristic of Down's syndrome.

65. S. J. Tredwell, D. F. Smith, P. J. Macleod, and B. J. Wood, "Cervical spine anomalies in fetal alcohol syndrome," *Spine* 7, no. 4 (1982): 331.

66. J. M. Garber, "Steep corneal curvature: a fetal alcohol syndrome landmark," *Journal of the American Optometric Association* 32, no. 6 (1984): 595.

67. Jones et al., "Pattern of malformation in offspring," 1267–71; Jones and Smith, "Recognition of fetal alcohol syndrome," 999–1001; Jones et al., "Outcome in offspring," 1076–78.

68. See, e.g., Hanson et al., 1458–60; Sterling K. Clarren and David W. Smith, "The fetal alcohol syndrome," *New England Journal of Medicine* 298, no. 19 (1978): 1063–67; Gary G. Gordon and Charles S. Lieber, "The fetal alcohol syndrome: mechanisms and models" (editorial), *Hospital Practice* 14, no. 10 (1979): 11, 15; National Institute of Child Health and Human Development (NICHD), "Fetal-alcohol syndrome," *Pediatric Annals* 8, no. 2 (1979): 106–7; Smith, "Fetal alcohol syndrome" 121–28; Sterling K. Clarren, "Recognition of fetal alcohol syndrome," *Journal of the American Medical Association* 245, no. 23 (1981): 2436–39.

69. Ann P. Streissguth, "Maternal drinking and the outcome of pregnancy: implications of child mental health," *American Journal of Orthopsychiatry* 3, no. 47 (1977): 422–31.

70. Ann P. Streissguth, "Fetal alcohol syndrome: an epidemiologic perspective," *American Journal of Epidemiology* 107, no. 6 (1978): 467–78.

71. Smith, "Fetal alcohol syndrome."

72. Kenneth Lyon Jones, "The fetal alcohol syndrome," *Addictive Diseases: An International Journal* 1–2, no. 2 (1975): 79–88.

73. Best.

74. Claire Toutant and Steven Lippmann, "Fetal alcohol syndrome," *American Family Physician* 1, no. 22 (1980): 114.

75. The Institute of Medicine's 1996 report on FAS is a sterling example of this kind of diagnosis expansion. It created a complex hierarchy of degrees of FAS, including a designation "partial FAS." What such a term truly means — that the observed birth defects are themselves only "partial" ones or that alcohol only "partially" caused the defects — remains unclear.

76. Gillian Turner, "The fetal alcohol syndrome," *Medical Journal of Australia* 1, no. 65 (1978): 18.

77. Smith, "Fetal alcohol syndrome," 121.

78. Ibid.

79. Food and Drug Administration.

80. Best, 58.

81. Nicholas J. Eastman and Louis M. Hellman, eds., *Williams Obstetrics*, 12th ed. (New York: Appleton-Century-Crofts, 1961), 348.

82. Louis M. Hellman and Jack A. Pritchard, eds., *Williams Obstetrics*, 14th ed. (New York: Appleton-Century-Crofts, 1971), 342

83. Ibid.

84. Jack A. Pritchard and Paul C. McDonald, eds., *Williams Obstetrics*, 15th ed. (New York: Appleton-Century-Crofts, 1976).

85. Ibid., 258.

86. Jack A. Pritchard and Paul C. McDonald, eds., *Williams Obstetrics*, 16th ed. (New York: Appleton-Century-Crofts, 1980).

87. David N. Danforth, ed., *Textbook of Obstetrics and Gynecology*, 2d ed. (New York: Harper and Row, Harper Medical Division, 1971).

88. David N. Danforth, ed., *Textbook of Obstetrics and Gynecology*, 3d ed. (New York: Harper and Row, Harper Medical Division, 1977), 557, 621.

89. Claudia Holzman and Nigel Paneth, "Preterm birth: from prediction to prevention," *American Journal of Public Health* 88, no. 2 (1998): 183–84; Robert L. Goldenberg, Jay D. Iams, Brian M. Mercer, Paul J. Meis, Atef H. Moawad, Rachel L. Copper, Anita Das, Elizabeth Thom, Francee Johnson, Donald McNellis, Menachem Miodovnik, J. Peter Van Dorsten, Steve N. Caritis, Gary R. Thurnau, Sidney F. Bottoms, and the NICHD MFMU Network, "The Preterm Prediction Study: the value of new vs. standard risk factors in predicting early and all spontaneous preterm births," *American Journal of Public Health* 88, no. 2 (1998): 233–38.

90. David N. Danforth, ed., *Textbook of Obstetrics and Gynecology*, 4th ed. (New York: Harper and Row, Harper Medical Division, 1982), 371.

91. Ibid.

92. Ibid., 208.

93. David N. Danforth, ed., *Textbook of Obstetrics and Gynecology*, 5th ed. (New York: Harper and Row, Harper Medical Division, 1986), 373.

94. Ibid., 359.

95. Ibid., 686.

96. Leslie Iffy and Harold A. Kaminetzky, eds., *Principles and Practices of Obstetrics and Perinatology* (New York: John Wiley, 1981), 1461.

97. Niswander, 232.

98. Jones et al., "Pattern of malformation," 1267.

99. Ibid., 1270.

100. Jones and Smith, "Recognition of fetal alcohol syndrome," 999.

101. Lynn Erb and Brian D. Andresen, "The fetal alcohol syndrome (FAS): a review of the impact of chronic maternal alcoholism on the developing fetus," *Clinical Pediatrics* 8, no. 17 (1978): 644.

102. Clarren and Smith, "Fetal alcohol syndrome," *New England Journal of Medicine*, 1063; Smith, "Fetal alcohol syndrome," 121.

103. Tenbrinck and Buchin, 1144; Clarren, 2436.

104. Kenneth Lyon Jones and David W. Smith, "The fetal alcohol syndrome," *Teratology* 12 (1975): 1.

105. Mulvihill et al., 941; Warner and Rosett, 1395–1420; L. W. Buckalew, "Alcohol: a description and comparison of recent scientific vs. public knowledge," *Journal of Clinical Psychology* 35, no. 2 (1979): 459–63.

106. Jones and Smith, "Recognition of fetal alcohol syndrome," 999–1001; H. Gordon Green, "Infants of alcoholic mothers," *American Journal of Obstetrics and Gynecology* 118, no. 5 (1974): 713–16; Jones and Smith, "Fetal alcohol syndrome," 1–10; Gilbert E. Corrigan, "The fetal alcohol syndrome," *Texas Medicine* 72, no. 1 (1976): 72–4; G. Turner.

107. Green, 713.

108. Abel, "Consumption of alcohol."

109. Krous, 295.

110. Abel, "Consumption of alcohol."

111. Jones and Smith, "Recognition of fetal alcohol syndrome"; Jones and Smith, "Fetal alcohol syndrome"; Streissguth, "Maternal drinking"; Streissguth, "Fetal alcohol syndrome"; G. Turner; Smith, "Fetal alcohol syndrome"; Elana Lindor, Anne-Marie McCarthy, and Maureen Guarino McRae, "Fetal alcohol syndrome: a review and case presentation," *Journal of Obstetric, Gynecologic, and Neonatal Nursing* 4, no. 9 (1980): 222–28; Wilson.

112. Clarren and Smith, "Fetal alcohol syndrome," *New England Journal of Medicine*; NICHD; Smith, "Fetal alcohol syndrome"; Krous; Wilson; Abel, "Consumption of alcohol."

113. Clarren and Smith, "Fetal alcohol syndrome," *New England Journal of Medicine*, 1063.

114. Abel, *Fetal Alcohol Syndrome*, 1.

115. Moira Plant, *Women and Alcohol: Contemporary and Historical Perspectives* (New York: Free Association Books, 1997).

116. Robert Graves, *The Greek Myths*, vol. 1 (New York: Penguin Books, 1955); Edith Hamilton, *Mythology* (New York: Penguin Books, 1969).

117. Abel, *Fetal Alcohol Syndrome*, 1.

118. Mary Douglas, *Purity and Danger: An Analysis of the Concepts of Pollution and Taboo* (London: Routledge, 1966).

119. Jones and Smith, "Recognition of fetal alcohol syndrome"; Jones and Smith, "Fetal alcohol syndrome," 1–10; Streissguth, "Maternal drinking"; Streissguth, "Fetal alcohol syndrome"; G. Turner; Smith, "Fetal alcohol syndrome."

120. Jones and Smith, "Recognition of fetal alcohol syndrome"; Green; Jones and Smith, "Fetal alcohol syndrome"; Streissguth, "Maternal drinking"; Streissguth, "Fetal alcohol syndrome"; Wilson.

121. Alvin E. Rodin, "Infants and gin mania in 18th century London," *Journal of the American Medical Association* 245, no. 12 (1981): 1239.

122. Henry L. Rosett, "Effects of maternal drinking on child development: an introductory review," *Annals of the New York Academy of Sciences* 273 (1976): 116.

123. Streissguth, "Fetal alcohol syndrome."

124. Ibid.

125. Streissguth, "Maternal drinking"; Lindor et al.

126. G. Turner.

127. Lindor et al.

128. Abel, *Fetal Alcohol Syndrome*.

129. Jones and Smith, "Recognition of fetal alcohol syndrome."

130. Rodin, 1238.

131. Green, 713.

132. NICHD, 106.

133. In fact, many doctors I interviewed (see Chapter 4), reflecting on their own or their parents' experiences of drinking during pregnancy, were somewhat incredulous that "social drinking" led to birth defects.

134. Carrie L. Randall, "Introduction to Fetal Alcohol Syndrome," in *Currents in Alcoholism*, vol. 5, *Biomedical Issues and Clinical Effects of Alcoholism*, ed. M. Galanter (New York: Grune and Stratton, 1979), 121.

135. Lindor et al., 222.

136. Best, 17.

137. Krous, 307.

138. Joseph Gusfield, "The literary rhetoric of science: comedy and pathos in drinking driver research," *American Sociological Review* 41 (1976): 16–34.

139. Jones and Smith, "Recognition of fetal alcohol syndrome."

140. G. Obe and F. Majewski, "No elevation of exchange type aberrations in lymphocytes of children with alcohol embryopathy," *Human Genetics* 43, no. 1 (1978): 31–36.

141. Sneed, 324.

142. Abel, *Fetal Alcohol Abuse Syndrome*.

143. Robert Merton "Priorities in Scientific Discovery," in *The Sociology of Science: Theoretical and Empirical Investigations* (Chicago: University of Chicago Press, 1973).

144. G. Turner; Chernoff; Willena S. Beagle, "Fetal alcohol syndrome: a review," *Journal of the American Dietetic Association* 79 (1981): 274–76; Clarren; Q. H. Qazi, A. Chua, D. Milman, and G. Solish, "Factors influencing outcome of pregnancy in heavy-drinking women," *Developmental Pharmacology and Therapeutics* 1–2, no. 4 (1982): 6–11; Patricia A. McCarthy, "Fetal alcohol syndrome and other alcohol-related birth defects," *Nurse Practitioner* 8, no. 1 (1983): 33–37; Ernest L. Abel, "Prenatal effects of alcohol," *Drug and Alcohol Dependence* 14, no. 1 (1984): 1–10; Kathleen Stratton, Cynthia Howe, and Frederick Battaglia, eds., *Fetal Alcohol Syndrome: Diagnosis, Epidemiology, Prevention, and Treatment* (Washington, D.C.: National Academy Press, 1996). Note that in *Threatened Children*, Best documented a similar progression from "battered child syndrome" to "child abuse."

145. Mendelson, 556.

146. J. J. Powell, "The tragedy of fetal alcohol syndrome," *RN* 44, no. 12 (1981): 33–35.

147. J. Beattie, "Fetal alcohol syndrome — the incurable hangover," *Health Visitor* 54, no. 11 (1981): 468–69.

148. W. F. Stanage, J. B. Gregg, and L. J. Massa, "Fetal alcohol syndrome — intrauterine child abuse," *South Dakota Journal of Medicine* 36, no. 10 (1983): 35.

149. Gusfield, *Culture of Public Problems*, 74.

150. Corrigan, 73.

151. Ibid.

152. Green; James W. Hanson, Ann P. Streissguth, and David W. Smith, "The effects of moderate alcohol consumption during pregnancy on fetal growth and morphogenesis," *Journal of Pediatrics* 92, no. 3 (1978): 457–60.

153. Jennie Kline, Patrick Shrout, Zena Stein, Mervyn Susser, and Dorothy Warburton, "Drinking during pregnancy and spontaneous abortion," *Lancet* 2, no. 8187 (1980): 176.

154. Ruth E. Little and Ann P. Streissguth, "Effects of alcohol on the fetus: impact and prevention," *Canadian Medical Association Journal* 125, no. 2 (1981): 159–64.

155. Lindor et al., 222–28; Robert J. Sokol, "Alcohol and abnormal outcomes of pregnancy," *Canadian Medical Association Journal* 125, no. 2 (1981): 143–48.

156. Krous, 306.

157. Ibid., 301.

158. Hanson et al.; Smith, "Fetal alcohol syndrome"; Janet Haggerty Davis and Wendy Autumn Frost, "Fetal alcohol syndrome: a challenge for the community health nurse," *Journal of Community Health Nursing* 1, no. 2 (1984): 99–110.

159. Smith, "Fetal alcohol syndrome," 124.

160. Erb and Andresen, 644.

161. Best, 41.

162. Stratton et al.

163. Best, 40.

164. Kathryn Montgomery Hunter, *Doctors' Stories: The Narrative Structure of Medical Knowledge* (Princeton: Princeton University Press, 1991).

165. Ibid., 28.

166. Best, 29.

167. Kenneth Lyon Jones and David W. Smith, "Offspring of chronic alcoholic women" (reply to letter to the editor), *Lancet* 2, no. 7876 (1974): 349.

168. Katherine K. Christoffel and Ira Salafsky, "Fetal alcohol syndrome in dizygotic twins," *Journal of Pediatrics* 87, no. 6 (1975): 967.

169. Jones et al., "Outcome in offspring," 1078.

Chapter 4. Charting Uncertainty through Doctors' Lenses

1. Peter L. Berger and Thomas Luckmann, *The Social Construction of Reality* (New York: Doubleday, 1966), 65.

2. This social investment is evident in such programs as the WIC (Women, Infants, and Children) nutritional supplement program for pregnant and nursing women and

the expanded Medicaid eligibility for pregnant women. As underfunded and inadequate as these social welfare programs are, that they exist at all suggests the extent to which we as a society regard pregnancy and reproduction as worthy of social attention and investment.

3. The sample was a convenience sample. I had contacts in each of the three fields, and they recommended a list of their colleagues who might be willing and able to grant an interview. Thus, my respondents are not a random sample of doctors in these specialties but a selective group. Among the doctors that I contacted, the response rate was quite high (more than 90 percent). Most doctors who declined to be interviewed said they were too busy. One family practitioner in my initial sample did not see pregnant women or children in his practice.

4. Journal reading, however, is an important activity for all practicing physicians, and several of the doctors in my sample reported that they made a concerted effort to keep up with the journals they subscribed to.

5. American Medical Association (AMA), *Physician Characteristics and Distribution in the U.S., 1997–1998* (Chicago: AMA, 1998). According to the AMA, the population of all physicians (not just AMA members) in the United States in 1996 was as follows: all specialties: 79 percent men, 21 percent women; obstetrics and gynecology: 69 percent men, 31 percent women; pediatrics: 55 percent men, 45 percent women; general/family practice: 78 percent men, 22 percent women.

6. Development is not yet a board-certified subspecialty of pediatrics, although there is a movement to make it one.

7. Several doctors in this sample reported spending some time on American Indian reservations. However, not all of them reported seeing FAS there.

8. Michael Dorris, *The Broken Cord* (New York: Harper Perennial, 1989).

9. Ibid., 264.

10. Ibid., xiii.

11. Kathryn Montgomery Hunter, *Doctors' Stories: The Narrative Structure of Medical Knowledge* (Princeton: Princeton University Press, 1991).

12. Joel Best, *Threatened Children: Rhetoric and Concern about Child-Victims* (Chicago: University of Chicago Press, 1990).

13. Ibid., 28.

14. David W. Smith, *Recognizable Patterns of Human Malformation: Genetic, Embryonic, and Clinical Aspects*, 2d ed. (Philadelphia: W. B. Saunders, 1976).

15. Hunter.

16. I was surprised that doctors would look to me, a sociologist with no medical training, as an authority on FAS. One doctor wanted me to tell her how to recognize FAS in her patients, information I felt I did not possess. She agreed to the interview on the condition that "if after you interview me, you tell me about FAS."

17. According to the CDC, the state of Pennsylvania has one of the highest percentages of frequent drinkers among women of childbearing age in the United States. Centers for Disease Control, "Frequent alcohol consumption among women of child-

bearing age — Behavioral Risk Factor Surveillance System, 1991," *Morbidity and Mortality Weekly Report* 43, no. 18 (13 May 1994): 328–29, 335.

18. Christopher Lawrence, "'Definite and Material': Coronary Thrombosis and Cardiologists in the 1920s," in *Framing Disease*, ed. Charles E. Rosenberg and Janet Golden (New Brunswick, N.J.: Rutgers University Press, 1992).

19. Lynn Payer, *Medicine and Culture* (New York: Penguin Books, 1988).

20. Talcott Parsons, "Social Structure and Dynamic Process: The Case of Modern Medical Practice," in *The Social System* (Glencoe, Ill.: Free Press, 1951), 456.

21. Kathleen Stratton, Cynthia Howe, and Frederick Battaglia, eds., *Fetal Alcohol Syndrome: Diagnosis, Epidemiology, Prevention, and Treatment* (Washington, D.C.: National Academy Press, 1996).

22. *Gold standard* is a term widely invoked by doctors to describe the generic problem of diagnosis, not just diagnosis of FAS.

23. Stephen J. Pfohl, "The 'discovery' of child abuse," *Social Problems* 24 (1977): 310–23.

24. Marcia Russell, "New assessment tools for risk drinking during pregnancy: T-ACE, TWEAK, and others," *Alcohol Health and Research World* 18, no. 1 (1994): 55.

25. This sentiment echoes something of late-nineteenth-century ideas about the hereditary effect of alcohol as a cause of degeneration in families.

26. Patients who are "drunks" are in fact highly stigmatized within medicine today. Many doctors (though by no means all) are loath to treat the recurrent medical conditions of alcoholics and find their behavior morally objectionable.

27. Renée C. Fox, "The Medicalization and Demedicalization of American Society," in *Essays in Medical Sociology* (New Brunswick, N.J.: Transaction Books, 1988), 467.

28. Cynthia R. Daniels, "Between fathers and fetuses: the social construction of male reproduction and the politics of fetal harm," *Signs* 22, no. 3 (1997): 583.

29. Peter Conrad, "Wellness as virtue: morality and the pursuit of health," *Culture, Medicine, and Psychiatry* 18 (1994): 385–401.

30. Parsons, 433.

31. Jane Balin, "The sacred dimensions of pregnancy and birth," *Qualitative Sociology* 11, no. 4 (1988): 275–301.

32. Since I began working on this topic, friends and acquaintances have bombarded me with anecdotes about heavy-handed instances of social control they have experienced while pregnant. These range from a liquor store owner telling a pregnant friend that it was "illegal" for him to sell her a very expensive bottle of cognac, to a convenience store clerk refusing to sell a colleague a pack of gum because she was pregnant. Almost inevitably when I mention my book topic, two things occur: I am asked whether I have read Michael Dorris's *The Broken Cord* and I am told a story about an outrageous incident that occurred during a pregnancy.

33. Best, 3.

34. Daniels, 583.

35. Parsons; Renée C. Fox, "The Evolution of Medical Uncertainty," reprinted in her *Essays in Medical Sociology.*

36. Fox, "Evolution of Medical Uncertainty," 567.

37. Centers for Disease Control, "Temporal trends in the incidence of birth defects—United States," *Morbidity and Mortality Weekly Report* 46, no. 49 (12 Dec. 1997): 1171–76.

38. William Ray Arney, *Power and the Profession of Obstetrics* (Chicago: University of Chicago Press, 1982), 139.

39. Ibid., 142.

40. Fox, "Evolution of Medical Uncertainty," 535.

41. Ana V. Diez-Roux, "Bringing context back into epidemiology: variables and fallacies in multilevel analysis," *American Journal of Public Health* 88, no. 2 (1998): 216–22.

42. Arney, 142.

43. Beverly Winikoff, *A Reassessment of the Concept of Reproductive Risk in Maternity Care and Family Planning Services,* proceedings of a seminar presented under the Population Council's Robert H. Ebert Program on Critical Issues in Reproductive Health and Population (New York: Population Council, 1991).

44. Daniels, 583.

45. Folate prevents neural tube defects.

46. Bruce G. Link and Jo Phelan, "Social conditions as fundamental causes of disease," *Journal of Health and Social Behavior,* extra issue (1995): 80–94.

47. The Main Line represents the affluent suburbs north of Philadelphia.

48. Berger and Luckmann, 19.

49. Best.

50. Diana Scully, *Men Who Control Women's Health: The Miseducation of Obstetrician-Gynecologists* (Boston: Houghton-Mifflin, 1980).

51. Joseph R. Gusfield, *The Culture of Public Problems: Drinking-Driving and the Social Order* (Chicago: University of Chicago Press, 1981), 63.

52. Ibid., 9.

Chapter 5. Discordant Depictions of Risk

1. Kathleen Stratton, Cynthia Howe, and Frederick Battaglia, eds., *Fetal Alcohol Syndrome: Diagnosis, Epidemiology, Prevention, and Treatment* (Washington, D.C.: National Academy Press, 1996), 63.

2. Jon M. Aase, "Clinical recognition of FAS: difficulties of detection and diagnosis," *Alcohol Health and Research World* 18, no. 1 (1994): 8.

3. Stratton et al.; Aase.

4. Susan J. Astley and Sterling K. Clarren, "Diagnosing the full spectrum of fetal alcohol-exposed individuals: introducing the 4-digit diagnostic code," *Alcohol and Alcoholism* 35, no. 4 (2000): 400.

5. Stratton et al.; José F. Cordero, R. Louise Floyd, M. Louise Martin, Margarett

Davis, and Karen Hymbaugh, "Tracking the prevalence of FAS," *Alcohol Health and Research World* 18, no. 1 (1994): 82–85.

6. Astley and Clarren.

7. Susan J. Astley, Sterling K. Clarren, M. Gratzer, A. Orkand, and M. Astion, *Fetal Alcohol Syndrome Tutor*[TM] *Medical Training Software*, CD-ROM (Washington, D.C.: March of Dimes, 1999).

8. Aase, 9.

9. Ibid.

10. Stratton et al., 63; Cordero et al.; Centers for Disease Control (CDC), "Fetal alcohol syndrome — United States, 1979–1992," *Morbidity and Mortality Weekly Report* 42, no. 17 (7 May 1993): 339–41; Centers for Disease Control, "Surveillance for fetal alcohol syndrome using multiple sources — Atlanta, Georgia, 1981–1989," *Morbidity and Mortality Weekly Report* 46, no. 47 (28 Nov. 1997): 1118–20.

11. CDC, "Surveillance for fetal alcohol syndrome."

12. J. C. King and S. Fabro, "Alcohol consumption and cigarette smoking: effect on pregnancy," *Clinical Obstetrics and Gynecology* 26 (1983): 437–48.

13. Aase.

14. Ibid.

15. Centers for Disease Control, "Birth certificate as a source for fetal alcohol syndrome case ascertainment — Georgia, 1989–1992," *Morbidity and Mortality Weekly Report* 44, no. 13 (7 Apr. 1995): 251–53.

16. CDC, "Fetal alcohol syndrome — United States, 1979–1992." The BDMP does not collect data from all fifty states, so the absolute number does not apply to the United States as a whole.

17. Stephanie J. Ventura, "Advance report of new data from the 1989 birth certificate," *Monthly Vital Statistics Report* 40, no. 12, suppl. (1992).

18. Centers for Disease Control, "Linking multiple data sources in fetal alcohol syndrome surveillance — Alaska," *Morbidity and Mortality Weekly Report* 42, no. 16 (30 Apr. 1993): 312–14.

19. CDC, "Surveillance for fetal alcohol syndrome."

20. CDC, "Birth certificate."

21. Philip A. May, Karen J. Hymbaugh, Jon M. Aase, and Jonathan M. Samet, "Epidemiology of fetal alcohol syndrome among American Indians of the Southwest," *Social Biology* 30, no. 4 (1983): 374–87.

22. Cindy Duimstra, Darlene Johnson, Charlotte Kutsch, Belle Wang, Miriam Zentner, Scott Kellerman, and Thomas Welty, "A fetal alcohol syndrome surveillance pilot project in American Indian communities in the Northern Plains," *Public Health Reports* 108, no. 2 (1993): 225–29.

23. Paul D. Sampson, Ann P. Streissguth, Fred L. Bookstein, Ruth E. Little, Sterling K. Clarren, Phillippe Dehaene, James W. Hanson, and John M. Graham, "Incidence of fetal alcohol syndrome and prevalence of alcohol-related neurodevelopmental disorder," *Teratology* 56 (1997): 317–26.

24. Linda S. Orgain and Madeline J. Caporale, "Using local media to educate

young women about FAS and FAE," *Public Health Reports* 108, no. 2 (Mar.-Apr. 1993): 171.

25. Hoover Adger and Mark J. Werner, "The pediatrician," *Alcohol Health and Research World* 18, no. 2 (1994): 121–26.

26. "Rate of alcohol-injured newborns soars," *Chicago Tribune*, 7 Apr. 1995, N6.

27. "Use of alcohol linked to rise in fetal illness," *New York Times*, 7 Apr. 1995.

28. Centers for Disease Control, "Update: trends in fetal alcohol syndrome — United States, 1979–1993," *Morbidity and Mortality Weekly Report* 44, no. 13 (7 Apr. 1995): 249–51.

29. Cordero et al.

30. Gilberto F. Chávez, José F. Cordero, and José E. Becerra, "Leading major congenital malformations among minority groups in the United States, 1981–1986," *Morbidity and Mortality Weekly Report* 37, no. SS-03 (1 July 1988): 17–24.

31. Jane Schulman, Larry D. Edmonds, Anne B. McClearn, Nancy Jensvold, and Gary M. Shaw, "Surveillance for and comparison of birth defect prevalences in two geographic areas — United States, 1983–88," *Morbidity and Mortality Weekly Report* 42, no. SS-1 (19 Mar. 1993): 1–6.

32. Best, 29.

33. Ibid.

34. Joseph Gusfield, *The Culture of Public Problems: Drinking-Driving and the Symbolic Order* (Chicago: University of Chicago Press, 1981).

35. Ibid., 58.

36. Best, 58.

37. Ibid., 45.

38. A typical drink is defined as a glass of wine, a beer, or a mixed drink. Each of these contains the same amount of absolute alcohol (ethanol), 0.5 ounces.

39. Robert J. Sokol, Susan S. Martier, and Joel W. Ager, "The T-ACE questions: practical prenatal detection of risk-drinking," *American Journal of Obstetrics and Gynecology* 160, no. 4 (Apr. 1989): 863–70.

40. The 1985 National Health Interview Survey found that among women of all ages, 59 percent reported no alcohol consumption in the two weeks prior to the survey and only 3 percent reported drinking two or more drinks per day (Gerald D. Williams, Mary Dufour, and Darryl Bertolucci, "Drinking levels, knowledge, and associated characteristics, 1985 NHIS findings," *Public Health Reports* 101, no. 6 [Nov.–Dec. 1986]: 593–98). A national survey of women aged 21 and older in 1991 found that about a quarter of women aged 21 to 34 and about a third of women aged 35 to 49 do not drink at all. Less than 5 percent in each age group reported heavy drinking (Sharon C. Wilsnack, Richard W. Wilsnack, and Susanne Hiller-Sturmhöfel, "How women drink: epidemiology of women's drinking and problem drinking," *Alcohol Health and Research World* 18, no. 3 [1994]: 173–81).

41. Centers for Disease Control, "Alcohol consumption among pregnant and childbearing-aged women — United States, 1991 and 1995," *Morbidity and Mortality Weekly Report* 46, no. 16 (25 Apr. 1997): 346–50.

42. Stephanie J. Ventura, Selma M. Taffel, and T. J. Mathews, "Advance report of maternal and infant health data from the birth certificate, 1990," *Monthly Vital Statistics Report* 42, suppl. (1993).

43. Daniel H. Ershoff, Neil K. Aaronson, Brian G. Danaher, and Fred W. Wasserman, "Behavioral, health and cost outcomes of an HMO-based prenatal health education program," *Public Health Reports* 98, no. 6 (1983): 536–47.

44. Sokol et al., "T-ACE questions."

45. Norman M. Bradburn and Seymour Sudman, *Polls and Surveys: Understanding What They Tell Us* (San Francisco: Jossey-Bass, 1988); Centers for Disease Control, "Frequent alcohol consumption among women of childbearing age — Behavioral Risk Factor Surveillance System, 1991," *Morbidity and Mortality Weekly Report* 43, no. 18 (13 May 1994): 328–29, 335; Centers for Disease Control, "Sociodemographic and behavioral characteristics associated with alcohol consumption during pregnancy — United States, 1988," *Morbidity and Mortality Weekly Report* 44, no. 13 (7 Apr. 1995): 261–64; Paul H. Verkerk, "The impact of alcohol misclassification on the relationship between alcohol and pregnancy outcome," *International Journal of Epidemiology* 21, no. 4, suppl. 1 (1992): S33–S37.

46. N. L. Fox, M. J. Sexton, and J. Richard Hebel, "Alcohol consumption among pregnant smokers: effects of a smoking cessation intervention program," *American Journal of Public Health* 77, no. 2 (1987): 211–13.

47. Joyce C. Abma and Frank L. Mott, "Is There a 'Bad Mother' Syndrome? An Analysis of Overlapping High Risk Factors during Pregnancy" (paper presented at the Annual Meeting of the Population Association of America, 1990).

48. C. A. Hilton and J. T. Condon, "Changes in smoking and drinking during pregnancy," *Australian and New Zealand Journal of Obstetrics and Gynaecology* 29, no. 1 (1989): 18–21; J. T. Condon and C. A. Hilton, "A comparison of smoking and drinking behaviors in pregnant women: who abstains and why," *Medical Journal of Australia* 148 (1988): 381–85; Jerry Kruse, M. LeFevre, and S. Zweig, "Changes in smoking and alcohol consumption during pregnancy: a population-based study in a rural area," *Obstetrics and Gynecology* 67 (1986): 627–32.

49. Kruse et al.

50. J. T. Wright, I. G. Barrison, I. G. Lewis, K. D. MacRae, E. J. Waterson, P. J. Toplis, M. G. Gordon, and N. F. Morris, "Alcohol consumption, pregnancy, and low birthweight," *Lancet* 1, no. 8326 (1983): 663–67; J. W. Kuzma and D. G. Kissinger, "Patterns of alcohol and cigarette use in pregnancy," *Neurobehavioral Toxicology and Teratology* 3, no. 2 (1981): 211–21; Fox et al., 211–13.

51. Ben G. Armstrong, Alison D. McDonald, and Margaret Sloan, "Cigarette, alcohol and coffee consumption and spontaneous abortion," *American Journal of Public Health* 82, no. 1 (1992): 85–87; I. G. Barrison and J. T. Wright, "Moderate drinking during pregnancy and fetal outcome," *Alcohol and Alcoholism* 19, no. 2 (1984): 167–72; King and Fabro, "Alcohol consumption and cigarette smoking"; Robert J. Sokol, S. I. Miller, S. Debanne, N. Golden, G. Collins, J. Kaplan, and Susan Martier, "The Cleveland NIAAA prospective alcohol-in-pregnancy study: the first year," *Neurobehavioral*

Toxicology and Teratology 3, no. 2 (1981): 203–9; J. Kline, P. Shrout, Z. Stein, M. Susser, and D. Warburton, "Drinking during pregnancy and spontaneous abortion," *Lancet* 2, no. 8187 (1980): 176–80.

52. Raul Caetano, "Drinking and alcohol-related problems among minority women," *Alcohol Health and Research World* 3, no. 18 (1994): 233–41.

53. Abma and Mott; J. Lumley, J. F. Correy, N. M. Newman, and J. T. Curran, "Cigarette smoking, alcohol consumption and fetal outcome in Tasmania 1981–82," *Australian and New Zealand Journal of Obstetrics and Gynaecology* 25 (1985): 33–40; P. Kwok, J. F. Correy, N. M. Newman, and J. T. Curran, "Smoking and alcohol consumption during pregnancy: an epidemiological study in Tasmania," *Medical Journal of Australia* 1 (1983): 220–23; Wright et al.; Kate Prager, Henry Malin, Danielle Spiegler, Pearl Van Natta, and Paul J. Placek, "Smoking and drinking behavior before and during pregnancy of married mothers of live-born infants and stillborn infants," *Public Health Reports* 99, no. 2 (Mar.-Apr. 1984): 117–27; Kuzma and Kissinger.

54. Abma and Mott.

55. Prager et al.

56. Mary Serdula, David F. Williamson, Juliette S. Kendrick, Robert F. Anda, and Tim Byers, "Trends in alcohol consumption by pregnant women, 1985 through 1988," *Journal of the American Medical Association* 265, no. 7 (1991): 876–79.

57. Bradburn and Sudman; CDC, "Frequent alcohol consumption"; CDC, "Sociodemographic and behavioral characteristics"; Verkerk.

58. The somewhat lower proportions of women in the NMIHS (compared with other national samples) who reported drinking may be partly explained by the inclusion in this sample of teenage women, who are less likely to drink.

59. However, women who experience an unintended pregnancy tend to be younger, poorer, and less educated. These characteristics are also associated with lower rates of drinking.

60. Some of the categories have been combined to avoid small cell sizes.

61. Michael D. Kogan, Milton Kotelchuck, Greg R. Alexander, and W. E. Johnson, "Racial disparities in reported prenatal care advice from health care providers," *American Journal of Public Health* 84, no. 1 (1994): 82–88.

62. Ernest L. Abel and Robert J. Sokol, "Incidence of fetal alcohol syndrome and economic impact of FAS-related anomalies," *Drug and Alcohol Dependency* 19, no. 1 (Jan. 1987): 51–70.

63. Kenneth R. Warren and Richard J. Bast, "Alcohol-related birth defects: an update," *Public Health Reports* 103, no. 6 (Nov.-Dec. 1988): 638–42; M. Plant, F. M. Sullivan, C. Guerri, and E. L. Abel, "Alcohol and Pregnancy," in *Health Issues Related to Alcohol Consumption*, ed. Paulus M. Verschuren (Brussels: International Life Sciences Institute, 1993); Ian Walpole, Stephen Zubrick, and Jacqueline Pontré, "Confounding variables in studying the effects of maternal alcohol consumption before and during pregnancy," *Journal of Epidemiology and Community Health* 43 (1989): 153–61.

64. Ernest L. Abel and John H. Hannigan, "Maternal risk factors in fetal alcohol

syndrome: provocative and permissive influences," *Neurotoxicology and Teratology* 17, no. 4 (1995), 445–62

65. Lumley et al.

66. Walpole et al.

67. This emotionally weighted term is sometimes used in the literature on FAS. For examples, see E. J. Waterson and Iain M. Murray-Lyon, "Preventing alcohol-related birth damage: a review," *Social Science and Medicine* 30, no. 3 (1990): 349–64; May et al.

68. This tension is reflected in the debate about whether a confirmed history of maternal alcohol use is necessary to make the diagnosis or whether suspected FAS justifies further probing for maternal history. In other words, do we look at history of maternal drinking as the starting point or the endpoint of the diagnosis?

69. Philip A. May, "Fetal alcohol effects among North American Indians," *Alcohol Health and Research World* 15, no. 3 (1991): 239–48.

70. Gusfield, 55.

71. Cynthia Mayer, "When Indian culture, outside rules clash," *Philadelphia Inquirer*, 26 Nov. 1993, A1.

72. This journal was renamed *Alcohol Research and Health* in 1999.

73. Adger and Werner.

74. One wonders if we might also create a diagnosis of "fetal poverty effect" or "poverty-related birth defects" to describe the far more common negative birth outcomes routinely experienced by poor women.

75. National Organization on Fetal Alcohol Syndrome, www.nofas.org/ (2002).

76. Gusfield, 60.

77. Centers for Disease Control, "Temporal trends in the incidence of birth defects — United States," *Morbidity and Mortality Weekly Report* 46, no. 49 (12 Dec. 1997): 1171–76.

78. K. Caruso and R. ten Bensel, "Fetal alcohol syndrome and fetal alcohol effects: the University of Minnesota experience," *Minnesota Medicine* 76 (1993): 25.

79. Partnership to Prevent Fetal Alcohol Syndrome, "FAS Facts," http://prevention .samhsa.gov/faspartners/facts/facts (accessed 1 Aug. 2002).

80. Mike Lowry, foreword, in *The Challenge of Fetal Alcohol Syndrome: Overcoming Secondary Disabilities*, ed. Ann Streissguth and Jonathan Kanter (Seattle: University of Washington Press, 1997), vii.

81. A genetic registry established among Orthodox Jews to lower the incidence of common genetic diseases in their community provides a striking example of this social anxiety. Anxiety runs so deep that in some cases people are willing to act at great personal cost to avoid the potential birth of a child with a genetic disorder. Several people are reported to have broken off engagements or stopped dating when they learned their partner carried a recessive gene. Gina Kolata, "Nightmare or the dream of a new era in genetics?" *New York Times*, 7 Dec. 1993.

82. Thomas Prugh, "Point-of-purchase health warning notices," *Alcohol Health*

and Research World 10, no. 4 (1986): 36–37; Lynn M. Paltrow, David S. Cohen, and Corinne A. Carey, *Year 2000 Overview: Governmental Responses to Pregnant Women Who Use Alcohol or Other Drugs* (New York: National Advocates for Pregnant Women and the Women's Law Project, 2000); Substance Abuse and Mental Health Services Administration, "Point-of-Sale Signage Laws," http://fascenter.samhsa.gov/resource/signage.cfm (accessed 16 July 2003).

83. Lee Ann Kaskutas, "Interpretations of risk: the use of scientific information in the development of the alcohol warning label," *International Journal of Addictions* 30, no. 12 (1995): 1519–48; Janet R. Hankin, James J. Sloan, Ira J. Firestone, Joel W. Ager, Robert J. Sokol, and Susan S. Martier, "A time series analysis of the impact of the alcohol warning label on antenatal drinking," *Alcoholism: Clinical and Experimental Research* 17, no. 2 (Mar.-Apr. 1993): 284–89; Janet R. Hankin, Ira J. Firestone, James J. Sloan, Joel W. Ager, Allen C. Goodman, Robert J. Sokol, and Susan S. Martier, "The impact of the alcohol warning label on drinking during pregnancy," *Journal of Public Policy and Marketing* 12, no. 1 (spring 1993): 10–18; Janet R. Hankin, Ira J. Firestone, James J. Sloan, Joel W. Ager, Robert J. Sokol, and Susan S. Martier, "Time series analyses reveal differential impact of the alcohol warning label by drinking level," *Applied Behavioral Science Review* 2, no. 1 (1994): 47–59; Janet R. Hankin, James J. Sloan, Ira J. Firestone, Joel W. Ager, Robert J. Sokol, and Susan S. Martier, "Has awareness of the alcohol warning label reached its upper limit?" *Alcoholism: Clinical and Experimental Research* 20, no. 3 (1996): 440–44; Janet R. Hankin, Ira J. Firestone, James J. Sloan, Joel W. Ager, Robert J. Sokol, and Susan S. Martier, "Heeding the alcoholic beverage warning label during pregnancy: multiparae versus nulliparae," *Journal of Studies on Alcohol* 57, no. 2 (1996): 171–77; Janet R. Hankin, James J. Sloan, Ira J. Firestone, et al., "The alcohol beverage warning label: when did knowledge increase?" *Alcoholism: Clinical and Experimental Research* 17, no. 2 (Mar.-Apr. 1993): 428–30.

84. Hankin et al., "Has awareness reached its upper limit?"; Hankin et al., "Impact of the alcohol warning label"; Hankin et al., "Time series analysis."

85. Hankin et al., "Time series analysis."

86. Hankin et al., "Alcohol beverage warning label."

Chapter 6. Medical-Moral Authority and the Redefinition of Risk in the Twentieth Century

1. Robert A. Aronowitz, "Lyme disease: the social construction of a new disease and its social consequences," *Milbank Quarterly* 69, no. 1 (1991): 79.

2. Rebecca H. Warner and Henry L. Rosett, "The effects of drinking on offspring: an historical survey of the literature," *Journal of Studies on Alcohol* 36 (1975): 1397.

3. Edward J. Burger, "Health as a surrogate for the environment," *Daedalus* 119, no. 4 (1990): 133–53.

4. Phillip Knightley, Harold Evans, Elaine Potter, and Marjorie Wallace, *Suffer the Children: The Story of Thalidomide* (New York: Viking Press, 1979).

5. Ibid.

6. Ashley Montagu, *Life before Birth* (New York: Signet, 1965), 89.

7. C. H. Kempe, H. K. Silver, F. N. Silverman, W. Droegemu, and B. F. Steele, "The battered child syndrome," *Journal of the American Medical Association* 181, no. 1 (1962): 17–24.

8. Josh Greenfeld, "Wounded before birth: a father sadly pursues his adopted son's affliction," *Chicago Tribune*, 23 July 1989, C1.

9. Anne Steacy, "Bruised before birth: alcoholic mothers damage their babies," *Maclean's* 102 (14 Aug. 1989): 48.

10. Julio O. Apolo, "Child abuse in the unborn fetus," *International Pediatrics* 10 (1995): 214–17.

11. J. L. Holmgren, "Legal accountability and fetal alcohol syndrome: when fixing the blame doesn't fix the problem," *South Dakota Law Review* 36 (1991): 81–103.

12. Centers for Disease Control, "Temporal trends in the incidence of birth defects — United States," *Morbidity and Mortality Weekly Report* 46, no. 49 (12 Dec. 1997): 1171–76.

13. Rayna Rapp, *Testing Women, Testing the Fetus: The Social Impact of Amniocentesis in America* (London: Routledge, 1999).

14. Jane Menken, "Age and fertility: how late can you wait?" *Demography* 22, no. 4 (1985): 469–83.

15. Sam Preston, "Changing values and falling birth rates," *Population and Development Review* 12, suppl. (1986): 176–95; Nathan Keyfitz, "The family that does not reproduce itself," *Population and Development Review* 12, suppl. (1986): 139–54; Viviana A. Zelizer, *Pricing the Priceless Child: The Changing Social Value of Children* (Princeton: Princeton University Press, 1985); Menken, 469–83.

16. Robert Blank, *Regulating Reproduction* (New York: Columbia University Press, 1990); Glenn McGee, *The Perfect Baby: A Pragmatic Approach to Genetics* (Lanham, Md.: Rowman and Littlefield, 1997).

17. Cynthia R. Daniels, *At Women's Expense: State Power and the Politics of Fetal Rights* (Cambridge: Harvard University Press, 1993); Stephen Maynard-Moody, *The Dilemma of the Fetus: Fetal Research, Medical Progress, and Moral Politics* (New York: St. Martin's Press, 1995); Monica J. Casper, *The Making of the Unborn Patient: A Social Anatomy of Fetal Surgery* (New Brunswick, N.J.: Rutgers University Press, 1998).

18. Karen Newman, *Fetal Positions: Individualism, Science, Visuality* (Stanford, Calif.: Stanford University Press, 1996).

19. Daniels, 17.

20. Barbara Katz Rothman, *The Tentative Pregnancy: Prenatal Diagnosis and the Future of Motherhood* (New York: Penguin Books, 1986).

21. George J. Annas, "Pregnant women as fetal containers," *Hastings Center Report* 16, no. 6 (1986): 13–14.

22. Daniels, 1.

23. Kristin Luker, *Abortion and the Politics of Motherhood* (Berkeley: University of California Press, 1984), 193.

24. Concern for changing social mores was explicit in some of the early FAS liter-

ature; just as in the nineteenth century, writers often noted increased drinking among younger women as cause for concern (e.g., Ann P. Streissguth, "Maternal drinking and the outcome of pregnancy: implications of child mental health," *American Journal of Orthopsychiatry* 47, no. 3 [1977]: 422–31). "In the past, women who drank excessively tended to be in an older age group, usually past their childbearing period, but now drinking beer and wine is becoming part of the teenage way of life" (Gillian Turner, "The fetal alcohol syndrome," *Medical Journal of Australia* 65, no. 1 [1978]: 19).

25. Daniels, 3.

26. Sterling K. Clarren, "Recognition of fetal alcohol syndrome," *Journal of the American Medical Association* 245, no. 23 (1981): 2439.

27. Emily Martin, *The Woman in the Body* (Boston: Beacon Press, 1987), 20.

28. Michael Dorris, *The Broken Cord* (New York: Harper Perennial, 1989); Daniels, 17.

29. Sylvia Tesh, "Disease causality and politics," *Journal of Health Politics, Policy and Law* 6 (1981): 369–90.

30. Irving K. Zola, "Medicine as an institution of social control," *Sociological Review* 20 (1972): 500.

31. Willena S. Beagle, "Fetal alcohol syndrome: a review," *Journal of the American Dietetic Association* 79 (1981): 276.

32. Sterling K. Clarren and David W. Smith, "The fetal alcohol syndrome," *New England Journal of Medicine* 298, no. 19 (1978): 556.

33. Kenneth L. Jones, David W. Smith, Ann P. Streissguth, and Ntinos C. Myrianthopoulos, "Outcome in offspring of chronic alcoholic women," *Lancet* 1, no. 7866 (1974): 1078.

34. Ann Streissguth, "The Ethanol Bath: The Fetal Alcohol and Drug Unit's Continuing Battle against Preventable Developmental Disorders," http://depts.washington.edu/psychweb/Newsletter/Spring2000/FADU.html (accessed 2 Aug. 2001).

35. ACOG statement of policy. This policy statement was withdrawn in September 1989. Currently, ACOG maintains that "Although an occasional drink during pregnancy has not been shown to be of harm, patients should be counseled that there is no level of alcohol use during pregnancy that is known to be safe." American College of Obstetricians and Gynecologists, *Substance Abuse in Pregnancy*, technical bulletin no. 195 (Washington, D.C.: ACOG, July 1994).

36. Food and Drug Administration, "Surgeon general's advisory on alcohol and pregnancy," *FDA Drug Bulletin* 11, no. 2 (1981): 9–10.

37. Elizabeth M. Armstrong and Ernest L. Abel, "Fetal alcohol syndrome: the origins of a moral panic," *Alcohol and Alcoholism* 35, no. 3 (2000): 276–82.

38. National Institute on Alcohol Abuse and Alcoholism (NIAAA), *U.S. Apparent Consumption of Alcoholic Beverages Based on State Sales, Taxation, or Receipt Data*, U.S. Alcohol Epidemiologic Data Reference Manual, vol. 1, 3d ed. (Bethesda, Md.: NIAAA, Oct. 1997).

39. Public Law 100-690, 27 USC 201–211.

40. Information Access Company, "Group forms to battle fetal alcohol syndrome," *Alcoholism and Drug Abuse Weekly* 3, no. 4 (1991).

41. National Organization on Fetal Alcohol Syndrome, www.nofas.org/ (2002).

42. Greg Dawson, "An innocent inherits the anguish of alcohol," *Orlando Sentinel Tribune*, 3 Feb. 1992, C1.

43. "Pregnancy, alcohol can be deadly mix," *Minneapolis Star-Tribune*, 26 Apr. 1992, A18.

44. Warren King, "Drinking devastating to unborn," *Seattle Times*, 16 Apr. 1991, B1.

45. Mike Snider, "Kids pay for prenatal drinking," *USA Today*, 10 Dec. 1990, 1D.

46. Dru Wilson, "Children pay the ultimate price for a drink," *Houston Chronicle* 11 Aug. 1998, 3.

47. "Prescription for tragedy: alcohol and pregnancy stack deck against baby," *Seattle Times*, 14 Jan. 1996, A17.

48. Phyllis Theroux, "The tragic inheritance: a father's chronicle of fetal alcohol syndrome," *Washington Post*, 19 July 1989, D3.

49. Lynn Paltrow, David S. Cohen, and Corinne A. Carey, *Year 2000 Overview: Governmental Responses to Pregnant Women Who Use Alcohol or Other Drugs* (New York: National Advocates for Pregnant Women and Women's Law Project, 2000); Jean Reith Schroedel and Pamela Fiber, "Punitive versus public health oriented responses to drug use by pregnant women," *Yale Journal of Health Policy, Law and Ethics* 1 (spring 2001): 217–35.

50. Ernest Abel and Robert J. Sokol, "Revised conservative estimate of the incidence of FAS and its economic impact," *Alcoholism: Clinical and Experimental Research* 15, no. 3 (1991): 514–24.

51. These are 1996 estimates. World Health Organization, *Global Status Report on Alcohol* (Geneva: World Health Organization, 1999).

52. Royal College of Obstetricians and Gynaecologists (RCOG), "Guideline No. 9: Alcohol Consumption in Pregnancy" (London: RCOG, Nov. 1996).

53. Canadian Centre on Substance Abuse (CCSA) National Working Group on Policy, *Fetal Alcohol Syndrome: An Issue of Child and Family Health* (Ottawa: CCSA, Oct. 1996).

54. Ibid.

55. World Health Organization, *WHO Primary Health Care Approaches for Prevention and Control of Congenital and Genetic Disorders*, report of a WHO meeting, Cairo, 6–8 Dec. 1999 (Geneva: World Health Organization, 2001).

56. C. Du, V. Florey, and D. Taylor, "Introduction," *International Journal of Epidemiology* 21, no. 4, suppl. 1 (1992): S6.

57. J. Olsen, "Recommendations," *International Journal of Epidemiology* 21, no. 4, suppl. 1 (1992): S82.

58. Ibid.

59. Ibid.

60. Sheila Jasanoff, "American exceptionalism and the political acknowledgment of risk," *Daedalus* 119, no. 4 (1990): 61–81.

61. Personal communication with Deborah Frank, Professor of Pediatrics, Boston University School of Medicine (Nov. 1996).

62. National Institute on Alcohol Abuse and Alcoholism, *Eighth Special Report to the U.S. Congress on Alcohol and Health*, NIH pub. no. 94-3699 (Bethesda, Md.: National Institutes of Health, 1993).

63. National Center for Health Statistics, *National Vital Statistics Reports* 48, no. 11 (1998).

64. National Highway Traffic Safety Administration (NHTSA), *Traffic Safety Facts, 1999*, DOT HS 809 086 (Washington, D.C.: NHTSA, n.d.).

65. Enoch Gordis, "Alcohol research and social policy: an overview," *Alcohol Health and Research World* 20, no. 4 (1996): 208–12.

66. National Institute on Alcohol Abuse and Alcoholism, "New advances in alcoholism treatment," *Alcohol Alert*, no. 49 (Oct. 2000).

67. National Highway Traffic Safety Administration, *Traffic Safety Facts 1995: State Alcohol Estimates* (Washington, D.C.: U.S. Department of Transportation, 1996).

68. Gordis.

69. National Institute on Alcohol Abuse and Alcoholism, "Alcohol and the workplace," *Alcohol Alert*, no. 44 (July 1999).

70. National Institute on Alcohol Abuse and Alcoholism, "Alcohol, violence, and aggression," *Alcohol Alert*, no. 38 (Oct. 1997).

71. Kaiser Family Foundation and National Center of Addiction and Substance Abuse at Columbia University, "Substance Use and Sexual Health among Teens and Young Adults in the U.S." (fact sheet, Feb. 2002).

72. National Institute on Alcohol Abuse and Alcoholism, "Are women more vulnerable to alcohol's effects?" *Alcohol Alert*, no. 46 (Dec. 1999).

73. Peter Conrad and Joseph W. Schneider, *Deviance and Medicalization: From Badness to Sickness* (St. Louis: C. V. Mosby, 1980).

74. Ibid.; Catherine K. Riessman and Constance A. Nathanson, "The Management of Reproduction: Social Construction of Risk and Responsibility," in *Applications of Social Science to Clinical Medicine and Health Policy*, ed. Linda H. Aiken and David Mechanic (New Brunswick, N.J.: Rutgers University Press, 1986); Charles E. Rosenberg, "Banishing risk: or the more things change the more they stay the same," *Perspectives in Biology and Medicine* 39, no. 1 (1995): 28.

75. Joseph R. Gusfield, *The Culture of Public Problems: Drinking-Driving and the Symbolic Order* (Chicago: University of Chicago Press, 1981), 53.

76. Howard S. Becker, *Outsiders: Studies in the Sociology of Deviance* (New York: Free Press, 1966), 147.

77. Carol Rivard, "The fetal alcohol syndrome," *Journal of School Health* 49, no. 2 (1979): 96–98.

78. Eliot Freidson, *Profession of Medicine: A Study of the Sociology of Applied Knowledge* (New York: Dodd Mead, 1970), 251.

Chapter 7. Bearing Responsibility

1. Ludwig Fleck, *Genesis and Development of a Scientific Fact*, trans. Fred Bradley and Thaddeus J. Trenn (Basel, Switzerland: Benno Schwabe, 1935; reprint, Chicago: University of Chicago Press, 1979).

2. Allan M. Brandt, "The cigarette, risk, and American culture," *Daedalus* 119, no. 4 (1990): 172.

3. Nancy Scheper-Hughes and Margaret M. Lock, "The mindful body: a prolegomenon to future work in medical anthropology," *Medical Anthropology Quarterly* 1, no. 1 (1987): 7.

4. Charles E. Rosenberg and Caroll Smith-Rosenberg, "The Female Animal: Medical and Biological Views of Women," in Charles E. Rosenberg, *No Other Gods: On Science and American Social Thought* (Baltimore: Johns Hopkins University Press, 1976).

5. Talcott Parsons, "Social Structure and Dynamic Process: The Case of Modern Medical Practice," in *The Social System* (Glencoe, Ill.: Free Press, 1951), 450–51.

6. Joseph R. Gusfield, *The Culture of Public Problems: Drinking-Driving and the Social Order* (Chicago: University of Chicago Press, 1981), 63.

7. Scheper-Hughes and Lock, 10.

8. Jacquelyn C. Campbell, "Abuse during pregnancy: progress, policy and potential," *American Journal of Public Health* 88, no. 2 (1998): 185–87.

9. Brandt, 172.

10. Ibid.

11. Christopher Hamlin, "Could you starve to death in England in 1839? The Chadwick-Farr controversy and the loss of the 'social' in public health," *American Journal of Public Health* 85, no. 6 (1995): 856–66; Bruce G. Link and Jo Phelan, "Social conditions as fundamental causes of disease," *Journal of Health and Social Behavior*, extra issue (1995): 80–94.

12. Ernest L. Abel, "An update on incidence of FAS: FAS is not an equal opportunity birth defect," *Neurotoxicology and Teratology* 17, no. 4 (1995): 437–43.

13. Ernest L. Abel and John H. Hannigan, "Maternal risk factors in fetal alcohol syndrome: provocative and permissive influences," *Neurotoxicology and Teratology* 17, no. 4 (1995): 445–62.

14. National Institute on Alcohol Abuse and Alcoholism, "Are women more vulnerable to alcohol's effects?" *Alcohol Alert*, no. 46 (Dec. 1999).

15. Dave Daley, "Racine case embodies debate over fetal rights," *Milwaukee Journal Sentinel*, 8 Sept. 1996.

16. *State of Wisconsin v Deborah J. Z.*, No. 96-2797-CR, Court of Appeals of Wisconsin, District Two (1999 Wisc App), LEXIS 581.

INDEX

CPSIA information can be obtained
at www.ICGtesting.com
Printed in the USA
BVOW06s0824220118
505959BV00001B/34/P